Linked Data

Evolving the Web into a Global Data Space

Synthesis Lectures on the Semantic Web: Theory and Technology

Editors

James Hendler, *Rensselaer Polytechnic Institute*
Frank van Harmelen, *Vrije Universiteit Amsterdam*

Whether you call it the Semantic Web, Linked Data, or Web 3.0, a new generation of Web technologies is offering major advances in the evolution of the World Wide Web. As the first generation of this technology transitions out of the laboratory, new research is exploring how the growing Web of Data will change our world. While topics such as ontology-building and logics remain vital, new areas such as the use of semantics in Web search, the linking and use of open data on the Web, and future applications that will be supported by these technologies are becoming important research areas in their own right. Whether they be scientists, engineers or practitioners, Web users increasingly need to understand not just the new technologies of the Semantic Web, but to understand the principles by which those technologies work, and the best practices for assembling systems that integrate the different languages, resources, and functionalities that will be important in keeping the Web the rapidly expanding, and constantly changing, information space that has changed our lives. Topics to be covered:

- Semantic Web Principles from linked-data to ontology design

- Key Semantic Web technologies and algorithms

- Semantic Search and language technologies

- The Emerging "Web of Data" and its use in industry, government and university applications

- Trust, Social networking and collaboration technologies for the Semantic Web

- The economics of Semantic Web application adoption and use

- Publishing and Science on the Semantic Web

- Semantic Web in health care and life sciences

Linked Data: Evolving the Web into a Global Data Space
Tom Heath and Christian Bizer
2011

Linked Data: Evolving the Web into a Global Data Space
Tom Heath and Christian Bizer

ISBN: 978-3-031-79431-5   paperback
ISBN: 978-3-031-79432-2   ebook

DOI 10.1007/978-3-031-79432-2

A Publication in the Springer series
SYNTHESIS LECTURES ON THE SEMANTIC WEB: THEORY AND TECHNOLOGY

Lecture #1
Series Editors: James Hendler, *Rensselaer Polytechnic Institute*
                         Frank van Harmelen, *Vrije Universiteit Amsterdam*
First Edition

10 9 8 7 6 5 4 3 2 1

Series ISSN
Synthesis Lectures on the Semantic Web: Theory and Technology
ISSN pending.

Linked Data

Evolving the Web into a Global Data Space

Tom Heath
Talis

Christian Bizer
Freie Universität Berlin

SYNTHESIS LECTURES ON THE SEMANTIC WEB: THEORY AND TECHNOLOGY #1

ABSTRACT

The World Wide Web has enabled the creation of a global information space comprising linked documents. As the Web becomes ever more enmeshed with our daily lives, there is a growing desire for direct access to raw data not currently available on the Web or bound up in hypertext documents. Linked Data provides a publishing paradigm in which not only documents, but also data, can be a first class citizen of the Web, thereby enabling the extension of the Web with a global data space based on open standards - the Web of Data. In this Synthesis lecture we provide readers with a detailed technical introduction to Linked Data. We begin by outlining the basic principles of Linked Data, including coverage of relevant aspects of Web architecture. The remainder of the text is based around two main themes - the publication and consumption of Linked Data. Drawing on a practical Linked Data scenario, we provide guidance and best practices on: architectural approaches to publishing Linked Data; choosing URIs and vocabularies to identify and describe resources; deciding what data to return in a description of a resource on the Web; methods and frameworks for automated linking of data sets; and testing and debugging approaches for Linked Data deployments. We give an overview of existing Linked Data applications and then examine the architectures that are used to consume Linked Data from the Web, alongside existing tools and frameworks that enable these. Readers can expect to gain a rich technical understanding of Linked Data fundamentals, as the basis for application development, research or further study.

KEYWORDS

web technology, databases, linked data, web of data, semantic web, world wide web, dataspaces, data integration, data management, web engineering, resource description framework

Contents

List of Figures

Preface

This book provides a conceptual and technical introduction to the field of Linked Data. It is intended for anyone who cares about data – using it, managing it, sharing it, interacting with it – and is passionate about the Web. We think this will include data geeks, managers and owners of data sets, system implementors and Web developers. We hope that students and teachers of information management and computer science will find the book a suitable reference point for courses that explore topics in Web development and data management. Established practitioners of Linked Data will find in this book a distillation of much of their knowledge and experience, and a reference work that can bring this to all those who follow in their footsteps.

Chapter 2 introduces the basic principles and terminology of Linked Data. Chapter 3 provides a 30,000 ft view of the Web of Data that has arisen from the publication of large volumes of Linked Data on the Web. Chapter 4 discusses the primary design considerations that must be taken into account when preparing to publish Linked Data, covering topics such as choosing and using URIs, describing things using RDF, data licensing and waivers, and linking data to external data sets. Chapter 5 introduces a number of recipes that highlight the wide variety of approaches that can be adopted to publish Linked Data, while Chapter 6 describes deployed Linked Data applications and examines their architecture. The book concludes in Chapter 7 with a summary and discussion of the outlook for Linked Data.

We would like to thank the series editors Jim Hendler and Frank van Harmelen for giving us the opportunity and the impetus to write this book. Summarizing the state of the art in Linked Data was a job that needed doing – we are glad they asked us. It has been a long process, throughout which Mike Morgan of Morgan & Claypool has shown the patience of a saint, for which we are extremely grateful. Richard Cyganiak wrote a significant portion of the 2007 tutorial "How to Publish Linked Data on the Web" which inspired a number of sections of this book – thank you Richard. Mike Bergman, Dan Brickley, Fabio Ciravegna, Ian Dickinson, John Goodwin, Harry Halpin, Frank van Harmelen, Olaf Hartig, Andreas Harth, Michael Hausenblas, Jim Hendler, Bernadette Hyland, Toby Inkster, Anja Jentzsch, Libby Miller, Yves Raimond, Matthew Rowe, Daniel Schwabe, Denny Vrandecic, and David Wood reviewed drafts of the book and provided valuable feedback when we needed fresh pairs of eyes – they deserve our gratitude. We also thank the European Commission for supporting the creation of this book by funding the LATC – LOD Around The Clock project (Ref. No. 256975). Lastly, we would like to thank the developers of LaTeX and Subversion, without which this exercise in remote, collaborative authoring would not have been possible.

Tom Heath and Christian Bizer
February 2011

CHAPTER 1

Introduction

1.1 THE DATA DELUGE

We are surrounded by data – data about the performance of our locals schools, the fuel efficiency of our cars, a multitude of products from different vendors, or the way our taxes are spent. By helping us make better decisions, this data is playing an increasingly central role in our lives and driving the emergence of a data economy [47]. Increasing numbers of individuals and organizations are contributing to this deluge by choosing to share their data with others, including *Web-native* companies such as *Amazon* and *Yahoo!*, newspapers such as *The Guardian* and *The New York Times*, public bodies such as the UK and US governments, and research initiatives within various scientific disciplines.

Third parties, in turn, are consuming this data to build new businesses, streamline online commerce, accelerate scientific progress, and enhance the democratic process. For example:

- The online retailer Amazon makes their product data available to third parties via a Web *API*[1]. In doing so they have created a highly successful ecosystem of affiliates[2] who build micro-businesses, based on driving transactions to Amazon sites.

- Search engines such as Google and Yahoo! consume structured data from the Web sites of various online stores, and use this to enhance the search listings of items from these stores. Users and online retailers benefit through enhanced user experience and higher transaction rates, while the search engines need expend fewer resources on extracting structured data from plain HTML pages.

- Innovation in disciplines such as Life Sciences requires the world-wide exchange of research data between scientists, as demonstrated by the progress resulting from cooperative initiatives such as the Human Genome Project.

- The availability of data about the political process, such as members of parliament, voting records, and transcripts of debates, has enabled the organisation *mySociety*[3] to create services such as *TheyWorkForYou*[4], through which voters can readily assess the performance of elected representatives.

---

[1] API stands for *Application Programming Interface* - a mechanism for enabling interaction between different software programs.
[2] https://affiliate-program.amazon.co.uk/
[3] http://www.mysociety.org/
[4] http://www.theyworkforyou.com/

The strength and diversity of the ecosystems that have evolved in these cases demonstrates a previously unrecognised, and certainly unfulfilled, demand for access to data, and that those organizations and individuals who choose to share data stand to benefit from the emergence of these ecosystems. This raises three key questions:

- How best to provide access to data so it can be most easily reused?

- How to enable the discovery of relevant data within the multitude of available data sets?

- How to enable applications to integrate data from large numbers of formerly unknown data sources?

Just as the World Wide Web has revolutionized the way we connect and consume documents, so can it revolutionize the way we *discover*, *access*, *integrate* and *use* data. The Web is the ideal medium to enable these processes, due to its ubiquity, its distributed and scalable nature, and its mature, well-understood technology stack.

The topic of this book is on how a set of principles and technologies, known as *Linked Data*, harnesses the ethos and infrastructure of the Web to enable data sharing and reuse on a massive scale.

1.2 THE RATIONALE FOR LINKED DATA

In order to understand the concept and value of Linked Data, it is important to consider contemporary mechanisms for sharing and reusing data on the Web.

1.2.1 STRUCTURE ENABLES SOPHISTICATED PROCESSING

A key factor in the re-usability of data is the extent to which it is well *structured*. The more regular and well-defined the structure of the data the more easily people can create tools to reliably process it for reuse.

While most Web sites have some degree of structure, the language in which they are created, HTML, is oriented towards structuring textual documents rather than data. As data is intermingled into the surrounding text, it is hard for software applications to extract snippets of structured data from HTML pages.

To address this issue, a variety of *microformats*[5] have been invented. Microformats can be used to published structured data describing specific types of entities, such as people and organizations, events, reviews and ratings, through embedding of data in HTML pages. As microformats tightly specify how to embed data, applications can unambiguously extract the data from the pages. Weak points of microformats are that they are restricted to representing data about a small set of different types of entities; they only provide a small set of attributes that may used to describe these entities; and that it is often not possible to express relationships between entities, such as, for example, that

---

[5]http://microformats.org/

a person is the speaker of an event, rather than being just an attendee or the organizer of the event. Therefore, microformats are not suitable for sharing arbitrary data on the Web.

A more generic approach to making structured data available on the Web are *Web APIs*. Web APIs provide simple query access to structured data over the HTTP protocol. High profile examples of these APIs include the *Amazon Product Advertising API*[6] and the *Flickr API*[7]. The site *ProgrammableWeb*[8] maintains a directory containing several thousand Web APIs.

The advent of Web APIs has led to an explosion in small, specialized applications (or *mashups*) that combine data from several sources, each of which is accessed through an API specific to the data provider. While the benefits of programmatic access to structured data are indisputable, the existence of a specialized API for each data set creates a landscape where significant effort is required to integrate each novel data set into an application. Every programmer must understand the methods available to retrieve data from each API, and write custom code for accessing data from each data source.

1.2.2 HYPERLINKS CONNECT DISTRIBUTED DATA

It is common for Web APIs to provide results in structured data formats such as XML and JSON[9], which have extensive support in a wide range of programming languages. However, from a Web perspective, they have some limitations, which are best explained by comparison with HTML. The HTML specification defines the *anchor* element, a, one of the valid attributes of which is the href. When used together, the anchor tag and href attribute indicate an outgoing link from the current document. Web *user agents*, such as browsers and search engine crawlers, are programmed to recognize the significance of this combination, and either render a clickable link that a human user can follow, or to traverse the link directly in order to retrieve and process the referenced document. It is this connectivity between documents, supported by a standard syntax for indicating links, that has enabled the Web of documents. By contrast, the data returned from the majority of Web APIs does not have the equivalent of the HTML *anchor* tag and href attribute, to indicate links that should be followed to find related data.

Furthermore, many Web APIs refer to items of interest using identifiers that have only local scope – e.g., a product identifier *123456* that is meaningless when taken out of the context of that specific API. In such cases, there is no standard mechanism to refer to items described by one API in data returned by another.

Consequently, data returned from Web APIs typically exists as isolated fragments, lacking reliable onward links signposting the way to related data. Therefore, while Web APIs make data accessible *on the Web*, they do not place it truly *in the Web*, making it linkable and therefore discoverable.

---

[6]http://docs.amazonwebservices.com/AWSECommerceService/latest/DG/
[7]http://www.flickr.com/services/api/
[8]http://www.programmableweb.com/
[9]http://www.json.org/

To return to the comparison with HTML, the analogous situation would be a search engine that required *a priori* knowledge of all Web documents before it could assemble its index. To provide this *a priori* knowledge, every Web publisher would need to register each Web page with each search engine. The ability for anyone to add new documents to the Web at will, and for these documents to be automatically discovered by search engines and humans with browsers, have historically been key drivers of the Web's explosive growth. The same principles of linking, and therefore ease of discovery, can be applied to data on the Web, and Linked Data provides a technical solution to realize such linkage.

1.3   FROM DATA ISLANDS TO A GLOBAL DATA SPACE

Linking data distributed across the Web requires a standard mechanism for specifying the existence and meaning of connections between items described in this data.

This mechanism is provided by the Resource Description Framework (RDF), which is examined in detail in Chapter 2. The key things to note at this stage are that RDF provides a flexible way to describe things in the world – such as people, locations, or abstract concepts – and how they relate to other things. These statements of relationships between things are, in essence, links connecting things in the world. Therefore, if we wish to say that a book described in data from one API is for sale at a (physical) bookshop described in data from a second API, and that bookshop is located in a city described by data from a third, RDF enables us to do this, and publish this information on the Web in a form that others can discover and reuse.

To conclude the comparison with HTML documents and conventional Web APIs, the key features of RDF worth noting in this context are the following:

- **RDF links things, not just documents**: therefore, in the book selling example above, RDF links would not simply connect the data fragments from each API, but assert connections between the entities described in the data fragments – in this case the book, the bookshop and the city.

- **RDF links are typed**: HTML links typically indicate that two documents are related in some way, but mostly leave the user to infer the nature of the relationship. In contrast, RDF enables the data publisher to state explicitly the nature of the connection. Therefore, in practice, the links in the book selling example above would read something like: *mybook* `forSaleIn` *thatbookshop*, *thatbookshop* `locatedIn` *mycity*.

While these sorts of connections between things in the world may be implicit in XML or JSON data returned from Web APIs, RDF enables Web publishers to make these links explicit, and in such a way that RDF-aware applications can follow them to discover more data. Therefore, a Web in which data is both published and linked using RDF is a Web where data is significantly more discoverable, and therefore more usable.

Just as hyperlinks in the classic Web connect documents into a single global information space, Linked Data enables links to be set between items in different data sources and therefore connect these sources into a single global data space. The use of Web standards and a common data model make it possible to implement generic applications that operate over the complete data space. This is the essence of *Linked Data*.

Increasing numbers of data providers and application developers have adopted Linked Data. In doing so they have created this global, interconnected data space - *the Web of Data*. Echoing the diversity of the classic document Web, the Web of Data spans numerous topical domains, such as people, companies, films, music, locations, books and other publications, online communities, as well as an increasing volume of scientific and government data.

This *Web of Data* [30], also referred to as *Semantic Web* [21], presents a revolutionary opportunity for deriving insight and value from data. By enabling seamless connections between data sets, we can transform the way drugs are discovered, create rich pathways through diverse learning resources, spot previously unseen factors in road traffic accidents, and scrutinise more effectively the operation of our democratic systems.

The focus of this book is data sharing in the context of the public Web. However, the principles and techniques described can be equally well applied to data that exists behind a personal or corporate firewall, or that straddles the public and the private. For example, many aspects of Linked Data have been implemented in desktop computing environments through the *Semantic Desktop* initiative[10]. Similarly, these principles can be employed entirely behind the corporate firewall, to help ease the pain of data integration in enterprise environments [114]. The *Linking Open Drug Data* [68] initiative represents a hybrid scenario, where Linked Data is enabling commercial organizations to connect and integrate data they are willing to share with each other for the purposes of collaboration.

1.4   INTRODUCING BIG LYNX PRODUCTIONS

Throughout this book we will illustrate the principles and technical aspects of Linked Data with examples from a scenario involving *Big Lynx Productions*. *Big Lynx* is a (fictional) independent television production company specialising in wildlife documentaries, primarily produced under contract for major television networks in the UK. The company employs around 30 permanent staff, such as *Managing Director* Dave Smith, *Lead Cameraman* Matt Briggs, and *Webmaster* Nelly Jones, plus a large team of freelancers that evolves according to the needs of current contracts.

Big Lynx maintains its own Web site at http://biglynx.co.uk/ that contains:

- information about the company's goals and structure

- profiles of the permanent staff and of freelancers

- listings of vacancies for freelancers to work on specific contracts

- listings of productions that have been broadcast by the commissioning network

- a blog where staff post news items of interest to the television networks and/or freelancers

---
[10]http://www.semanticdesktop.org/

Information that changes rarely (such as the company overview) is published on the site as static HTML documents. Frequently changing information (such as listing of productions) is stored in a relational database and published to the Web site as HTML by a series of PHP scripts developed for the company. The company blog is based on a blogging platform developed in-house and forms part of the main *Big Lynx* site.

In the remainder of this book we will explore how Linked Data can be integrated into the workflows and technical architectures of *Big Lynx*, thereby maximising the discoverability of the *Big Lynx* data and making it easy for search engines as well as specialized Web sites, such as film and TV sites, freelancer directories or online job markets, to pick up and integrate *Big Lynx* data with data from other companies.

CHAPTER 2

Principles of Linked Data

The term *Linked Data* refers to a set of best practices for publishing and interlinking structured data on the Web. These best practices were introduced by Tim Berners-Lee in his Web architecture note *Linked Data* [16] and have become known as the *Linked Data principles*. These principles are the following:

1. Use URIs as names for things.

2. Use HTTP URIs, so that people can look up those names.

3. When someone looks up a URI, provide useful information, using the standards (RDF, SPARQL).

4. Include links to other URIs, so that they can discover more things.

The basic idea of Linked Data is to apply the general architecture of the World Wide Web [67] to the task of sharing structured data on global scale. In order to understand these Linked Data principles, it is important to understand the architecture of the classic document Web.

The document Web is built on a small set of simple standards: Uniform Resource Identifiers (URIs) as globally unique identification mechanism [20], the Hypertext Transfer Protocol (HTTP) as universal access mechanism [50], and the Hypertext Markup Language (HTML) as a widely used content format [97]. In addition, the Web is built on the idea of setting hyperlinks between Web documents that may reside on different Web servers.

The development and use of standards enables the Web to transcend different technical architectures. Hyperlinks enable users to navigate between different servers. They also enable search engines to crawl the Web and to provide sophisticated search capabilities on top of crawled content. Hyperlinks are therefore crucial in connecting content from different servers into a *single global information space*. By combining simplicity with decentralization and openness, the Web seems to have hit an architectural sweet spot, as demonstrated by its rapid growth over the past 20 years.

Linked Data builds directly on Web architecture and applies this architecture to the task of sharing data on global scale.

2.1   THE PRINCIPLES IN A NUTSHELL

The first Linked Data principle advocates using URI references to identify, not just Web documents and digital content, but also real world objects and abstract concepts. These may include tangible things such as people, places and cars, or those that are more abstract, such as the relationship type of *knowing somebody*, the set of all green cars in the world, or the color green itself. This principle can be seen as extending the scope of the Web from online resources to encompass any object or concept in the world.

The HTTP protocol is the Web's universal access mechanism. In the classic Web, HTTP URIs are used to combine globally unique identification with a simple, well-understood retrieval mechanism. Thus, the second Linked Data principle advocates the use of HTTP URIs to identify objects and abstract concepts, enabling these URIs to be *dereferenced* (i.e., looked up) over the HTTP protocol into a description of the identified object or concept.

In order to enable a wide range of different applications to process Web content, it is important to agree on standardized content formats. The agreement on HTML as a dominant document format was an important factor that made the Web scale. The third Linked Data principle therefore advocates use of a single data model for publishing structured data on the Web – the Resource Description Framework (RDF), a simple graph-based data model that has been designed for use in the context of the Web [70]. The RDF data model is explained in more detail later in this chapter.

The fourth Linked Data principle advocates the use of hyperlinks to connect not only Web documents, but any type of thing. For example, a hyperlink may be set between a person and a place, or between a place and a company. In contrast to the classic Web where hyperlinks are largely untyped, hyperlinks that connect things in a Linked Data context have types which describe the relationship between the things. For example, a hyperlink of the type *friend of* may be set between two people, or a hyperlink of the type *based near* may be set between a person and a place. Hyperlinks in the Linked Data context are called *RDF links* in order to distinguish them from hyperlinks between classic Web documents.

Across the Web, many different servers are responsible for answering requests attempting to dereference HTTP URIs in many different namespaces, and (in a Linked Data context) returning RDF descriptions of the resources identified by these URIs. Therefore, in a Linked Data context, if an RDF link connects URIs in different namespaces, it ultimately connects resources in different data sets.

Just as hyperlinks in the classic Web connect documents into a single global information space, Linked Data uses hyperlinks to connect disparate data into a single global data space. These links, in turn, enable applications to navigate the data space. For example, a Linked Data application that has looked up a URI and retrieved RDF data describing a person may follow links from that data to data on different Web servers, describing, for instance, the place where the person lives or the company for which the person works.

As the resulting Web of Data is based on standards and a common data model, it becomes possible to implement generic applications that operate over the complete data space. Examples of such applications include Linked Data browsers which enable the user to view data from one data source and then follow RDF links within the data to other data sources. Other examples are Linked

Data Search engines that crawl the Web of Data and provide sophisticated query capabilities on top of the complete data space. Section 6.1 will give an overview of deployed Linked Data applications.

In summary, the Linked Data principles lay the foundations for extending the Web with a global data space based on the same architectural principles as the classic document Web. The following sections explain the technical realization of the Linked Data principles in more detail.

2.2   NAMING THINGS WITH URIS

To publish data on the Web, the items in a domain of interest must first be identified. These are the things whose properties and relationships will be described in the data, and may include Web documents as well as real-world entities and abstract concepts. As Linked Data builds directly on Web architecture [67], the Web architecture term *resource* is used to refer to these *things of interest*, which are, in turn, identified by HTTP URIs.

Figure 2.1[1] depicts the use of HTTP URIs to identify real-world entities and their rela-

Figure 2.1: URIs are used to identify people and the relationships between them.

tionships. The picture shows a *Big Lynx* film team at work. Within the picture, *Big Lynx* Lead Cameraman Matt Briggs as well as his two assistants, Linda Meyer and Scott Miller, are identified by HTTP URIs from the *Big Lynx* namespace. The relationship, that they know each other, is represented by connecting lines having the relationship type `http://xmlns.com/foaf/0.1/knows`.

[1]Please see copyright page for photo credits.

As discussed above, Linked Data uses only HTTP URIs, avoiding other URI schemes such as URNs [83] and DOIs [92]. HTTP URIs make good names for two reasons:

1. They provide a simple way to create globally unique names in a decentralized fashion, as every owner of a domain name, or delegate of the domain name owner, may create new URI references.

2. They serve not just as a name but also as a means of accessing information describing the identified entity.

If thinking about HTTP URIs as names for things rather than as addresses for Web documents feels strange to you, then references [113] and [106] are highly recommended reading and warrant re-visiting on a regular basis.

2.3   MAKING URIS DEFERERENCEABLE

Any HTTP URI should be dereferenceable, meaning that HTTP clients can look up the URI using the HTTP protocol and retrieve a description of the resource that is identified by the URI. This applies to URIs that are used to identify classic HTML documents, as well as URIs that are used in the Linked Data context to identify real-world objects and abstract concepts.

Descriptions of resources are embodied in the form of Web documents. Descriptions that are intended to be read by humans are often represented as HTML. Descriptions that are intended for consumption by machines are represented as RDF data.

Where URIs identify real-world objects, it is essential to not confuse the objects themselves with the Web documents that describe them. It is, therefore, common practice to use different URIs to identify the real-world object and the document that describes it, in order to be unambiguous. This practice allows separate statements to be made about an object and about a document that describes that object. For example, the creation date of a person may be rather different to the creation date of a document that describes this person. Being able to distinguish the two through use of different URIs is critical to the coherence of the Web of Data.

The Web is intended to be an information space that may be used by humans as well as by machines. Both should be able to retrieve representations of resources in a form that meets their needs, such as HTML for humans and RDF for machines. This can be achieved using an HTTP mechanism called *content negotiation* [50]. The basic idea of content negotiation is that HTTP clients send HTTP headers with each request to indicate what kinds of documents they prefer. Servers can inspect these headers and select an appropriate response. If the headers indicate that the client prefers HTML, then the server will respond by sending an HTML document. If the client prefers RDF, then the server will send the client an RDF document.

There are two different strategies to make URIs that identify real-world objects dereference-able. Both strategies ensure that objects and the documents that describe them are not confused, and that humans as well as machines can retrieve appropriate representations. The strategies are called *303 URIs* and *hash URIs*. The W3C Interest Group Note *Cool URIs for the Semantic Web* [98] describes and motivates both strategies in detail. The following sections summarize both strategies and illustrate each with an example HTTP session.

2.3.1   303 URIS

Real-world objects, like houses or people, can not be transmitted *over the wire* using the HTTP protocol. Thus, it is also not possible to directly dereference URIs that identify real-world objects. Therefore, in the *303 URIs* strategy, instead of sending the object itself over the network, the server responds to the client with the HTTP response code 303 See Other and the URI of a Web document which describes the real-world object. This is called a *303 redirect*. In a second step, the client dereferences this new URI and gets a Web document describing the real-world object.

Dereferencing a HTTP URI that identifies a real-world object or abstract concept thus involves a four step procedure:

1. The client performs a HTTP GET request on a URI identifying a real-world object or abstract concept. If the client is a Linked Data application and would prefer an RDF/XML representation of the resource, it sends an Accept: application/rdf+xml header along with the request. HTML browsers would send an Accept: text/html header instead.

2. The server recognizes that the URI identifies a real-world object or abstract concept. As the server can not return a representation of this resource, it answers using the HTTP 303 See Other response code and sends the client the URI of a Web document that describes the real-world object or abstract concept in the requested format.

3. The client now performs an HTTP GET request on this URI returned by the server.

4. The server answers with a HTTP response code 200 OK and sends the client the requested document, describing the original resource in the requested format.

This process can be illustrated with a concrete example. Imagine *Big Lynx* wants to serve data about their Managing Director Dave Smith on the Web. This data should be understandable for humans as well as for machines. *Big Lynx* therefore defines a URI reference that identifies the person Dave Smith (real-world object) and publishes two documents on its Web server: an RDF document containing the data about Dave Smith and an HTML document containing a human-readable representation of the same data. *Big Lynx* uses the following three URIs to refer to Dave and the two documents:

- http://biglynx.co.uk/people/dave-smith
  (URI identifying the person Dave Smith)

- http://biglynx.co.uk/people/dave-smith.rdf
  (URI identifying the RDF/XML document describing Dave Smith)

- http://biglynx.co.uk/people/dave-smith.html
  (URI identifying the HTML document describing Dave Smith)

To obtain the RDF data describing Dave Smith, a Linked Data client would connect to the http://biglynx.co.uk/ server and issue the following HTTP GET request:

```
1  GET /people/dave-smith HTTP/1.1
2  Host: biglynx.co.uk
3  Accept: text/html;q=0.5, application/rdf+xml
```

The client sends an `Accept:` header to indicate that it would take either HTML or RDF, but would prefer RDF. This preference is indicated by the *quality value* q=0.5 for HTML. The server would answer:

```
1  HTTP/1.1 303 See Other
2  Location: http://biglynx.co.uk/people/dave-smith.rdf
3  Vary: Accept
```

This is a 303 redirect, which tells the client that a Web document containing a description of the requested resource, in the requested format, can be found at the URI given in the `Location:` response header. Note that if the `Accept:` header had indicated a preference for HTML, the client would have been redirected to a different URI. This is indicated by the `Vary:` header, which is required so that caches work correctly. Next, the client will try to dereference the URI given in the response from the server.

```
1  GET /people/dave-smith.rdf HTTP/1.1
2  Host: biglynx.co.uk
3  Accept: text/html;q=0.5, application/rdf+xml
```

The *Big Lynx* Web server would answer this request by sending the client the RDF/XML document containing data about Dave Smith:

```
1  HTTP/1.1 200 OK
2  Content-Type: application/rdf+xml
3
4
5  <?xml version="1.0" encoding="UTF-8"?>
6  <rdf:RDF
7    xmlns:rdf="http://www.w3.org/1999/02/22-rdf-syntax-ns#"
8    xmlns:foaf="http://xmlns.com/foaf/0.1/">
9
10   <rdf:Description rdf:about="http://biglynx.co.uk/people/dave-smith">
11     <rdf:type rdf:resource="http://xmlns.com/foaf/0.1/Person"/>
12     <foaf:name>Dave Smith</foaf:name>
13     ...
```

The 200 status code tells the client that the response contains a representation of the requested resource. The `Content-Type:` header indicates that the representation is in RDF/XML format. The rest of the message contains the representation itself, in this case an RDF/XML description of Dave Smith. Only the beginning of this description is shown. The RDF data model, in general, will be described in 2.4.1, while the RDF/XML syntax itself will be described in Section 2.4.2.

2.3.2   HASH URIS

A widespread criticism of the *303 URI* strategy is that it requires two HTTP requests to retrieve a single description of a real-world object. One option for avoiding these two requests is provided by the *hash URI* strategy.

The *hash URI* strategy builds on the characteristic that URIs may contain a special part that is separated from the base part of the URI by a hash symbol (#). This special part is called the *fragment identifier*.

When a client wants to retrieve a hash URI, the HTTP protocol requires the fragment part to be stripped off before requesting the URI from the server. This means a URI that includes a hash cannot be retrieved directly and therefore does not necessarily identify a Web document. This enables such URIs to be used to identify real-world objects and abstract concepts, without creating ambiguity [98].

Big Lynx has defined various vocabulary terms in order to describe the company in data published on the Web. They may decide to use the hash URI strategy to serve an RDF/XML file containing the definitions of all these vocabulary terms. *Big Lynx* therefore assigns the URI

`http://biglynx.co.uk/vocab/sme`

to the file (which contains a vocabulary of Small and Medium-sized Enterprises) and appends *fragment identifiers* to the file's URI in order to identify the different vocabulary terms that are defined in the document. In this fashion, URIs such as the following are created for the vocabulary terms:

- `http://biglynx.co.uk/vocab/sme#SmallMediumEnterprise`

- `http://biglynx.co.uk/vocab/sme#Team`

To dereference any of these URIs, the HTTP communication between a client application and the server would look as follows:

First, the client truncates the URI, removing the fragment identifier (e.g., `#Team`). Then, it connects to the server at biglynx.co.uk and issues the following HTTP GET request:

```
1  GET /vocab/sme HTTP/1.1
2  Host: biglynx.co.uk
3  Accept: application/rdf+xml
```

The server answers by sending the requested RDF/XML document, an abbreviated version of which is shown below:

```
1  HTTP/1.1 200 OK
2  Content-Type: application/rdf+xml;charset=utf-8
3
4
5  <?xml version="1.0"?>
6  <rdf:RDF
7    xmlns:rdf="http://www.w3.org/1999/02/22-rdf-syntax-ns#"
8    xmlns:rdfs="http://www.w3.org/2000/01/rdf-schema#">
9
```

```
10    <rdf:Description
              rdf:about="http://biglynx.co.uk/vocab/sme#SmallMediumEnterprise">
11      <rdf:type rdf:resource="http://www.w3.org/2000/01/rdf-schema#Class" />
12    </rdf:Description>
13    <rdf:Description rdf:about="http://biglynx.co.uk/vocab/sme#Team">
14      <rdf:type rdf:resource="http://www.w3.org/2000/01/rdf-schema#Class" />
15    </rdf:Description>
16    ...
```

This demonstrates that the returned document contains not only a description of the vocabulary term http://biglynx.co.uk/vocab/sme#Team but also of the
term http://biglynx.co.uk/vocab/sme#SmallMediumEnterprise. The Linked Data-aware
client will now inspect the response and find triples that tell it more about the
http://biglynx.co.uk/vocab/sme#Team resource. If it is not interested in the triples describing the second resource, it can discard them before continuing to process the retrieved data.

2.3.3   HASH VERSUS 303

So which strategy should be used? Both approaches have their advantages and disadvantages. Section
4.4. of the W3C Interest Group Note *Cool URIs for the Semantic Web* compares both approaches [98]:
hash URIs have the advantage of reducing the number of necessary HTTP round-trips, which, in
turn, reduces access latency. The downside of the hash URI approach is that the descriptions of
all resources that share the same non-fragment URI part are always returned to the client together,
irrespective of whether the client is interested in only one URI or all. If these descriptions consist of a
large number of triples, the hash URI approach can lead to large amounts of data being unnecessarily
transmitted to the client. 303 URIs, on the other hand, are very flexible because the redirection target
can be configured separately for each resource. There could be one describing document for each
resource, or one large document for all of them, or any combination in between. It is also possible
to change the policy later on.

As a result of these factors, 303 URIs are often used to serve resource descriptions that are part
of very large data sets, such as the description of an individual concept from *DBpedia*, an RDF-ized
version of Wikipedia, consisting of 3.6 million concepts which are described by over 380 million
triples [32] (see Section 3.2.1 for a fuller description of DBpedia).

Hash URIs are often used to identify terms within RDF vocabularies, as the definitions of
RDF vocabularies are usually rather small, maybe a thousand RDF triples, and as it is also often
convenient for client applications to retrieve the complete vocabulary definition at once, instead of
having to look up every term separately. Hash URIs are also used when RDF is embedded into
HTML pages using RDFa (described in Section 2.4.2.2). Within the RDFa context, hash URIs are
defined using the RDFa about= attribute. Using them ensures that the URI of the HTML document
is not mixed up with the URIs of the resources described within this document.

It is also possible to combine the advantages of the 303 URI and the hash URI
approach. By using URIs that follow a http://domain/resource#this pattern, for instance,
http://biglynx.co.uk/vocab/sme/Team#this, you can flexibly configure what data is returned as

a description of a resource and still avoid the second HTTP request, as the #this part, which distinguished between the document and the described resource, is stripped off before the URI is dereferenced [98].

The examples in this book will use a mixture of Hash and 303 URIs to reflect the variety of usage in Linked Data published on the Web at large.

2.4   PROVIDING USEFUL RDF INFORMATION

In order to enable a wide range of different applications to process Web content, it is important to agree on standardized content formats. When publishing Linked Data on the Web, data is represented using the Resource Description Framework (RDF) [70]. RDF provides a data model that is extremely simple on the one hand but strictly tailored towards Web architecture on the other hand. To be published on the Web, RDF data can be serialized in different formats. The two RDF serialization formats most commonly used to published Linked Data on the Web are *RDF/XML* [9] and *RDFa* [1].

This section gives an overview of the RDF data model, followed by a comparison of the different RDF serialization formats that are used in the Linked Data context.

2.4.1   THE RDF DATA MODEL

The RDF data model [70] represents information as node-and-arc-labeled directed graphs. The data model is designed for the integrated representation of information that originates from multiple sources, is heterogeneously structured, and is represented using different schemata [12]. RDF aims at being employed as a *lingua franca*, capable of moderating between other data models that are used on the Web. The RDF data model is described in detail as part of the W3C RDF Primer [76]. Below, we give a short overview of the data model.

In RDF, a description of a resource is represented as a number of *triples*. The three parts of each triple are called its *subject*, *predicate*, and *object*. A triple mirrors the basic structure of a simple sentence, such as this one:

```
Matt Briggs   has nick name   Matty
   Subject       Predicate     Object
```

The subject of a triple is the URI identifying the described resource. The object can either be a simple *literal value*, like a string, number, or date; or the URI of another resource that is somehow related to the subject. The predicate, in the middle, indicates what kind of relation exists between subject and object, e.g., this is the name or date of birth (in the case of a literal), or the employer or someone the person knows (in the case of another resource). The predicate is also identified by a URI. These predicate URIs come from *vocabularies*, collections of URIs that can be used to represent information about a certain domain. Please refer to Section 4.4.4 for more information about which vocabularies to use in a Linked Data context.

Two principal types of RDF triples can be distinguished, *Literal Triples* and *RDF Links*:

1. **Literal Triples** have an RDF literal such as a string, number, or date as the object. Literal triples are used to describe the properties of resources. For instance, literal triples are used to describe the name or date of birth of a person. Literals may be plain or typed: A plain literal is a string combined with an optional language tag. The language tag identifies a natural language, such as English or German. A typed literal is a string combined with a datatype URI. The datatype URI identifies the datatype of the literal. Datatype URIs for common datatypes such as integers, floating point numbers and dates are defined by the XML Schema datatypes specification [26]. The first triple in the code example below is a literal triple, stating that *Big Lynx* Lead Cameraman Matt Briggs has the nick name *Matty*.

2. **RDF Links** describe the relationship between two resources. RDF links consist of three URI references. The URIs in the subject and the object position of the link identify the related resources. The URI in the predicate position defines the type of relationship between the resources. For instance, the second triple in the example below states that Matt Briggs *knows* Dave Smith. The third triple states that he leads something identified by the URI `http://biglynx.co.uk/teams/production` (in this case the *Big Lynx* Production Team). A useful distinction can be made between *internal* and *external* RDF links. *Internal RDF links* connect resources within a single Linked Data source. Thus, the subject and object URIs are in the same namespace. *External RDF links* connect resources that are served by different Linked Data sources. The subject and object URIs of external RDF links are in different namespaces. External RDF links are crucial for the Web of Data as they are the glue that connects data islands into a global, interconnected data space. The different roles that external RDF links have on the Web of Data will be discussed in detail in Section 2.5.

```
1   http://biglynx.co.uk/people/matt-briggs  http://xmlns.com/foaf/0.1/nick "Matty"
2   http://biglynx.co.uk/people/matt-briggs  http://xmlns.com/foaf/0.1/knows
        http://biglynx.co.uk/people/dave-smith
3   http://biglynx.co.uk/people/matt-briggs  http://biglynx.co.uk/vocab/sme#leads
        http://biglynx.co.uk/teams/production
```

One way to think of a set of RDF triples is as an RDF graph. The URIs occurring as subject and object are the nodes in the graph, and each triple is a directed arc that connects the subject and the object. As Linked Data URIs are globally unique and can be dereferenced into sets of RDF triples, it is possible to imagine all Linked Data as one *giant global graph*, as proposed by Tim Berners-Lee in [17]. Linked Data applications operate on top of this giant global graph and retrieve parts of it by dereferencing URIs as required.

2.4.1.1 Benefits of using the RDF Data Model in the Linked Data Context

The main benefits of using the RDF data model in a Linked Data context are that:

1. By using HTTP URIs as globally unique identifiers for data items as well as for vocabulary terms, the RDF data model is inherently designed for being used at global scale and enables anybody to refer to anything.

2. Clients can look up any URI in an RDF graph over the Web to retrieve additional information. Thus each RDF triple is part of the global Web of Data and each RDF triple can be used as a starting point to explore this data space.

3. The data model enables you to set RDF links between data from different sources.

4. Information from different sources can easily be combined by merging the two sets of triples into a single graph.

5. RDF allows you to represent information that is expressed using different schemata in a single graph, meaning that you can mix terms for different vocabularies to represent data. This practice is explained in Section 4.4.

6. Combined with schema languages such as RDF-Schema [37] and OWL [79], the data model allows the use of as much or as little structure as desired, meaning that tightly structured data as well as semi-structured data can be represented. A short introduction to RDF Schema and OWL is also given in Section 4.4.

2.4.1.2   RDF Features Best Avoided in the Linked Data Context

Besides the features mentioned above, the RDF Recommendation [70] also specifies a range of other features which have not achieved widespread adoption in the Linked Data community. In order to make it easier for clients to consume data, it is recommended to use only the subset of the RDF data model described above. In particular, the following features should be avoided in a Linked Data context.

1. **RDF reification** should be avoided, as reified statements are rather cumbersome to query with the SPARQL query language [95]. Instead of using reification to publish metadata about individual RDF statements, meta-information should instead be attached to the Web document containing the relevant triples, as explained in Section 4.3.

2. **RDF collections** and **RDF containers** are also problematic if the data needs to be queried with SPARQL. Therefore, in cases where the relative ordering of items in a set is not significant, the use of multiple triples with the same predicate is recommended.

3. The scope of **blank nodes** is limited to the document in which they appear, meaning it is not possible to create RDF links to them from external documents, reducing the potential for interlinking between different Linked Data sources. In addition, it becomes much more difficult to merge data from different sources when blank nodes are used, as there is no URI to serve as a common key. Therefore, all resources in a data set should be named using URI references.

2.4.2   RDF SERIALIZATION FORMATS

It is important to remember that RDF is not a data format, but a data model for describing resources in the form of *subject, predicate, object* triples. In order to publish an RDF graph on the Web, it must first be serialized using an RDF syntax. This simply means taking the triples that make up an RDF graph, and using a particular syntax to write these out to a file (either in advance for a static data set or on demand if the data set is more dynamic). Two RDF serialization formats - RDF/XML and RDFa - have been standardized by the W3C. In addition several other non-standard serialization formats are used to fulfill specific needs. The relative advantages and disadvantages of the different serialization formats are discussed below, along with a code sample showing a simple graph expressed in each serialization.

2.4.2.1 RDF/XML

The RDF/XML syntax [9] is standardized by the W3C and is widely used to publish Linked Data on the Web. However, the syntax is also viewed as difficult for humans to read and write, and, therefore, consideration should be given to using other serializations in data management and curation workflows that involve human intervention, and to the provision of alternative serializations for consumers who may wish to *eyeball* the data. The RDF/XML syntax is described in detail as part of the W3C RDF Primer [76]. The MIME type that should be used for RDF/XML within HTTP content negotiation is `application/rdf+xml`. The listing below show the RDF/XML serialization of two RDF triples. The first one states that there is a thing, identified by the URI `http://biglynx.co.uk/people/dave-smith` of type `Person`. The second triple state that this thing has the name Dave Smith.

```
1   <?xml version="1.0" encoding="UTF-8"?>
2   <rdf:RDF
3      xmlns:rdf="http://www.w3.org/1999/02/22-rdf-syntax-ns#"
4      xmlns:foaf="http://xmlns.com/foaf/0.1/">
5
6      <rdf:Description rdf:about="http://biglynx.co.uk/people/dave-smith">
7         <rdf:type rdf:resource="http://xmlns.com/foaf/0.1/Person"/>
8         <foaf:name>Dave Smith</foaf:name>
9      </rdf:Description>
10
11  </rdf:RDF>
```

2.4.2.2 RDFa

RDFa [1] is a serialization format that embeds RDF triples in HTML documents. The RDF data is not embedded in comments within the HTML document, as was the case with some early attempts to mix RDF and HTML, but is interwoven within the HTML *Document Object Model (DOM)*. This means that existing content within the page can be marked up with RDFa by modifying HTML code, thereby exposing structured data to the Web. A detailed introduction into RDFa is given in the W3C RDFa Primer [1].

RDFa is popular in contexts where data publishers are able to modify HTML templates but have relatively little additional control over the publishing infrastructure. For example, many content management systems will enable publishers to configure the HTML templates used to expose different types of information, but may not be flexible enough to support 303 redirects and HTTP content negotiation. When using RDFa to publish Linked Data on the Web, it is important to maintain the unambiguous distinction between the real-world objects described by the data and the HTML+RDFa document that embodies these descriptions. This can be achieved by using the RDFa about= attribute to assign URI references to the real-world objects described by the data. If these URIs are first defined in an RDFa document they will follow the *hash URI* pattern.

```
1   <!DOCTYPE html PUBLIC "−//W3C//DTD XHTML+RDFa 1.0//EN"
        "http://www.w3.org/MarkUp/DTD/xhtml−rdfa −1.dtd">
2   <html xmlns="http://www.w3.org/1999/xhtml"
        xmlns:rdf="http://www.w3.org/1999/02/22−rdf−syntax−ns#"
        xmlns:foaf="http://xmlns.com/foaf/0.1/">
3
4   <head>
5       <meta http−equiv="Content−Type" content="application/xhtml+xml;
            charset=UTF−8"/>
6       <title>Profile Page for Dave Smith</title>
7   </head>
8
9   <body>
10    <div about="http://biglynx.co.uk/people#dave−smith" typeof="foaf:Person">
11      <span property="foaf:name">Dave Smith</span>
12    </div>
13  </body>
14
15  </html>
```

2.4.2.3 Turtle

Turtle is a plain text format for serializing RDF data. Due to its support for namespace prefixes and various other shorthands, Turtle is typically the serialization format of choice for reading RDF triples or writing them by hand. A detailed introduction to Turtle is given in the W3C Team Submission document *Turtle - Terse RDF Triple Language* [10]. The MIME type for Turtle is `text/turtle;charset=utf-8`.

```
1   @prefix rdf: <http://www.w3.org/1999/02/22−rdf−syntax−ns#> .
2   @prefix foaf: <http://xmlns.com/foaf/0.1/> .
3
4   <http://biglynx.co.uk/people/dave−smith>
5     rdf:type foaf:Person ;
6     foaf:name "Dave Smith" .
```

2.4.2.4 N-Triples

N-Triples is a subset of Turtle, minus features such as namespace prefixes and shorthands. The result is a serialization format with lots of redundancy, as all URIs must be specified in full in each triple.

Consequently, N-Triples files can be rather large relative to Turtle and even RDF/XML. However, this redundancy is also the primary advantage of N-Triples over other serialization formats, as it enables N-Triples files to be parsed one line at a time, making it ideal for loading large data files that will not fit into main memory. The redundancy also makes N-Triples very amenable to compression, thereby reducing network traffic when exchanging files. These two factors make N-Triples the *de facto* standard for exchanging large dumps of Linked Data, e.g., for backup or mirroring purposes. The complete definition of the N-Triples syntax is given as part of the W3C *RDF Test Cases* recommendation[2].

```
1   <http://biglynx.co.uk/people/dave-smith>
        <http://www.w3.org/1999/02/22-rdf-syntax-ns#type>
        <http://xmlns.com/foaf/0.1/Person> .
2   <http://biglynx.co.uk/people/dave-smith> <http://xmlns.com/foaf/0.1/name> "Dave
    Smith" .
```

2.4.2.5 RDF/JSON

RDF/JSON refers to efforts to provide a *JSON* (JavaScript Object Notation) serialization for RDF, the most widely adopted of which is the Talis specification[3] [4]. Availability of a JSON serialization of RDF is highly desirable, as a growing number of programming languages provide native JSON support, including staples of Web programming such as *JavaScript* and *PHP*. Publishing RDF data in JSON therefore makes it accessible to Web developers without the need to install additional software libraries for parsing and manipulating RDF data. It is likely that further efforts will be made in the near future to standardize a JSON serialization of RDF[4].

2.5   INCLUDING LINKS TO OTHER THINGS

The forth Linked Data principle is to set RDF links pointing into other data sources on the Web. Such *external RDF links* are fundamental for the Web of Data as they are the glue that connects data islands into a global, interconnected data space and as they enable applications to discover additional data sources in a *follow-your-nose* fashion.

Technically, an external RDF link is an RDF triple in which the subject of the triple is a URI reference in the namespace of one data set, while the predicate and/or object of the triple are URI references pointing into the namespaces of other data sets. Dereferencing these URIs yields a description of the linked resource provided by the remote server. This description will usually contain additional RDF links which point to other URIs that, in turn, can also be dereferenced, and so on. This is how individual resource descriptions are woven into the Web of Data. This is also how the Web of Data can be navigated using a Linked Data browser or crawled by the robot of a search engine. There are three important types of RDF links:

---

[2]http://www.w3.org/TR/rdf-testcases/#ntriples
[3]http://n2.talis.com/wiki/RDF_JSON_Specification
[4]http://www.w3.org/2010/09/rdf-wg-charter.html

1. **Relationship Links** point at related things in other data sources, for instance, other people, places or genes. For example, relationship links enable people to point to background information about the place they live, or to bibliographic data about the publications they have written.

2. **Identity Links** point at URI aliases used by other data sources to identify the same real-world object or abstract concept. Identity links enable clients to retrieve further descriptions about an entity from other data sources. Identity links have an important social function as they enable different views of the world to be expressed on the Web of Data.

3. **Vocabulary Links** point from data to the definitions of the vocabulary terms that are used to represent the data, as well as from these definitions to the definitions of related terms in other vocabularies. Vocabulary links make data self-descriptive and enable Linked Data applications to understand and integrate data across vocabularies.

The following section gives examples of all three types of RDF link and discusses their role on the Web of Data.

2.5.1   RELATIONSHIP LINKS

The Web of Data contains information about a multitude of things ranging from people, companies, and places, to films, music, books, genes, and various other types of data. Chapter 3 will give an overview of the data sources that currently make up the Web of Data.

RDF links enable references to be set from within one data set to entities described in another, which may, in turn, have descriptions that refer to entities in a third data set, and so on. Therefore, setting RDF links not only connects one data source to another, but enables connections into a potentially infinite network of data that can be used collectively by client applications.

```
1   @prefix rdf: <http://www.w3.org/1999/02/22-rdf-syntax-ns#> .
2   @prefix foaf: <http://xmlns.com/foaf/0.1/> .
3
4   <http://biglynx.co.uk/people/dave-smith>
5     rdf:type foaf:Person ;
6     foaf:name "Dave Smith" ;
7     foaf:based_near <http://sws.geonames.org/3333125/> ;
8     foaf:based_near <http://dbpedia.org/resource/Birmingham> ;
9     foaf:topic_interest <http://dbpedia.org/resource/Wildlife_photography> ;
10    foaf:knows <http://dbpedia.org/resource/David_Attenborough> .
```

The example above demonstrates how *Big Lynx* uses RDF links pointing at related entities to enrich the data it publishes about its Managing Director Dave Smith. In order to provide background information about the place where he lives, the example contains an RDF link stating that Dave is `based_near` something identified by the URI `http://sws.geonames.org/3333125/`. Linked Data applications that look up this URI will retrieve a extensive description of Birmingham from Geonames[5], a data source that provides names of places (in different languages), geo-coordinates, and

[5]`http://www.geonames.org/`

information about administrative structures. The Geonames data about Birmingham will contain a further RDF link pointing at `http://dbpedia.org/resource/Birmingham`.

By following this link, applications can find population counts, postal codes, descriptions in 90 languages, and lists of famous people and bands that are related to Birmingham . The description of Birmingham provided by DBpedia, in turn, contains RDF links pointing at further data sources that contain data about Birmingham. Therefore, by setting a single RDF link, *Big Lynx* has enabled applications to retrieve data from a network of interlinked data sources.

2.5.2   IDENTITY LINKS

The fact that HTTP URIs are not only identifiers, but also a means to access information, results in many different URIs being used to refer to the same real-world object.

The rationale for, and implications of, this can be illustrated with an example of someone (who will be known as *Jeff*) who wants to publish data on the Web describing himself. Jeff must first define a URI to identify himself, in a namespace that he owns, or in which the domain name owner has allowed him to create new URIs. He then sets up a Web server to return the data describing himself, in response to someone looking up his URI over the HTTP protocol. After looking up the URI and receiving the descriptive data, an information consumer knows two things: first, the data about Jeff; second, the origin of that data, as he has retrieved the data from a URI under Jeff's control.

But what happens if Jeff wants to publish data describing a location or a famous person on the Web? The same procedure applies: Jeff defines URIs identifying the location and the famous person in his namespace and serves the data when somebody looks up these URIs. Information consumers that look up Jeff's URIs get his data and know again that he has published it.

In an open environment like the Web it is likely that Jeff is not the only one talking about the place or the famous person, but that there are many different information providers who talk about the same entities. As they all use their own URIs to refer to the person or place, the result is multiple URIs identifying the same entity. These URIs are called *URI aliases*.

In order to still be able to track the different information providers speak about the same entity, Linked Data relies on setting RDF links between URI aliases. By common agreement, Linked Data publishers use the link type `http://www.w3.org/2002/07/owl#sameAs` to state that two URI aliases refer to the same resource. For instance, if Dave Smith would also maintain a private data homepage besides the data that *Big Lynx* publishes about him, he could add a `http://www.w3.org/2002/07/owl#sameAs` link to his private data homepage, stating that the URI used to refer to him in this document and the URI used by *Big Lynx* both refer to the same real-world entity.

```
1   <http://www.dave-smith.eg.uk#me> <http://www.w3.org/2002/07/owl#sameAs>
        <http://biglynx.co.uk/people/dave-smith>  .
```

To use different URIs to refer to the same entity and to use `owl:sameAs` links to connect these URIs appears to be cumbersome at first sight, but is actually essential to make the Web of Data work as a social system. The reasons for this are:

1. **Different opinions.** URI aliases have an important social function on the Web of Data as they are dereferenced to descriptions of the same resource provided by different data publishers and thus allow different views and opinions to be expressed.

2. **Traceability.** Using different URIs allows consumers of Linked Data to know what a particular publisher has to say about a specific entity by dereferencing the URI that is used by this publisher to identify the entity.

3. **No central points of failure.** If all things in the world were to each have one, and only one, URI, this would entail the creation and operation of a centralized naming authority to assign URIs. The coordination complexity, administrative and bureaucratic overhead this would introduce would create a major barrier to growth in the Web of Data.

The last point becomes especially clear when one considers the size of many data sets that are part of the Web of Data. For instance, the Geonames data set provides information about over eight million locations. If in order to start publishing their data on the Web of Data, the Geonames team would need to find out what the commonly accepted URIs for all these places would be, doing so would be so much effort that it would likely prevent Geonames from publishing their dataset as Linked Data at all. Defining URIs for the locations in their own namespace lowers the barrier to entry, as they do not need to know about other people's URIs for these places. Later, they, or somebody else, may invest effort into finding and publishing `owl:sameAs` links pointing to data about these places other datasets, enabling progressive adoption of the Linked Data principles.

Therefore, in contrast to relying on upfront agreement on URIs, the Web of Linked Data relies on solving the identity resolution problem in an evolutionary and distributed fashion: evolutionary, in that more and more `owl:sameAs` links can be added over time; and distributed, in that different data providers can publish `owl:sameAs` links and as the overall effort for creating these links can thus shared between the different parties.

There has been significant uncertainty in recent years about whether `owl:sameAs` or other predicates should be used to express identity links [53]. A major source of this uncertainty is that the OWL semantics [93] treat RDF statements as facts rather then as claims by different information providers. Today, `owl:sameAs` is widely used in the Linked Data context and hundreds of millions of `owl:sameAs` links are published on the Web. Therefore, we recommend to also use `owl:sameAs` to express identity links, but always to keep in mind that the Web is a social system and that all its content needs to be treated as claims by different parties rather than as facts (see Section 6.3.5 on Data Quality Assessment). This guidance is also supported by members of the W3C Technical Architecture Group (TAG)[6].

---

[6]`http://lists.w3.org/Archives/Public/www-tag/2007Jul/0032.html`

2.5.3   VOCABULARY LINKS

The promise of the Web of Data is not only to enable client applications to discover new data sources by following RDF links at run-time but also to help them to integrate data from these sources. Integrating data requires bridging between the schemata that are used by different data sources to publish their data. The term *schema* is understood in the Linked Data context as the mixture of distinct terms from different RDF vocabularies that are used by a data source to publish data on the Web. This mixture may include terms from widely used vocabularies (see Section 4.4.4) as well as proprietary terms.

The Web of Data takes a two-fold approach to dealing with heterogeneous data representation [22]. On the one hand side, it tries to avoid heterogeneity by advocating the reuse of terms from widely deployed vocabularies. As discussed in Section 4.4.4 a set of vocabularies for describing common things like people, places or projects has emerged in the Linked Data community. Thus, whenever these vocabularies already contain the terms needed to represent a specific data set, they should be used. This helps to avoid heterogeneity by relying on ontological agreement.

On the other hand, the Web of Data tries to deal with heterogeneity by making data as self-descriptive as possible. Self-descriptiveness [80] means that a Linked Data application which discovers some data on the Web that is represented using a previously unknown vocabulary should be able to find all meta-information that it requires to translate the data into a representation that it understands and can process. Technically, this is realized in a twofold manner: first, by making the URIs that identify vocabulary terms dereferenceable so that client applications can look up the RDFS and OWL definition of terms – this means that every vocabulary term links to its own definition [23]; second, by publishing mappings between terms from different vocabularies in the form of RDF links [80]. Together these techniques enable Linked Data applications to discover the meta-information that they need to integrate data in a follow-your-nose fashion along RDF links.

Linked Data publishers should therefore adopt the following workflow: first, search for terms from widely used vocabularies that could be reused to represent data (as described in Section 4.4.4); if widely deployed vocabularies do not provide all terms that are needed to publish the complete content of a data set, the required terms should be defined as a proprietary vocabulary (as described in Section 4.4.6) and used in addition to terms from widely deployed vocabularies. Wherever possible, the publisher should seek wider adoption for the new, proprietary vocabulary from others with related data.

If at a later point in time, the data publisher discovers that another vocabulary contains the same term as the proprietary vocabulary, an RDF link should be set between the URIs identifying the two vocabulary terms, stating that these URIs actually refer to the same concept (= the term). The Web Ontology Language (OWL) [79], RDF Schema (RDFS) [37] and the Simple Knowledge Organization System (SKOS) [81] define RDF link types that can be used to represent such mappings. `owl:equivalentClass` and `owl:equivalentProperty` can be used to state that terms in different vocabularies are equivalent. If a looser mapping is desired, then `rdfs:subClassOf`, `rdfs:subPropertyOf`, `skos:broadMatch`, and `skos:narrowMatch` can be used.

The example below illustrates how the proprietary vocabulary term `http://biglynx.co.uk/vocab/sme#SmallMediumEnterprise` is interlinked with related terms from the DBpedia, Freebase, UMBEL, and OpenCyc.

```
1   @prefix rdf: <http://www.w3.org/1999/02/22-rdf-syntax-ns#> .
2   @prefix rdfs: <http://www.w3.org/2000/01/rdf-schema#> .
3   @prefix owl: <http://www.w3.org/2002/07/owl#> .
4   @prefix co: <http://biglynx.co.uk/vocab/sme#> .
5
6   <http://biglynx.co.uk/vocab/sme#SmallMediumEnterprise>
7     rdf:type rdfs:Class ;
8     rdfs:label "Small or Medium-sized Enterprise" ;
9     rdfs:subClassOf <http://dbpedia.org/ontology/Company> .
10    rdfs:subClassOf <http://umbel.org/umbel/sc/Business> ;
11    rdfs:subClassOf <http://sw.opencyc.org/concept/Mx4rvVjQNpwpEbGdrcN5Y29ycA> ;
12    rdfs:subClassOf <http://rdf.freebase.com/ns/m/0qb7t> .
```

Just as `owl:sameAs` links can be used to incrementally interconnect URI aliases, term-level links between different vocabularies can also be set over time by different parties. The more links that are set between vocabulary terms, the better client applications can integrate data that is represented using different vocabularies. Thus, the Web of Data relies on a distributed, *pay-as-you-go* approach to data integration, which enables the integration effort to be split over time and between different parties [51][74][34]. This type of data integration is discussed in more detail in Section 6.4.

2.6   CONCLUSIONS

This chapter has outlined the basic principles of Linked Data and has described how the principles interplay in order to extend the Web with a global data space. Similar to the classic document Web, the Web of Data is built on a small set of standards and the idea to use links to connect content from different sources. The extent of its dependence on URIs and HTTP demonstrates that Linked Data is not disjoint from the Web at large, but simply an application of its principles and key components to novel forms of usage. Far from being an additional layer on top of but separate from the Web, Linked Data is just another *warp* or *weft* being steadily interwoven with the fabric of the Web.

Structured data is made available on the Web today in forms. Data is published as CSV data dumps, Excel spreadsheets, and in a multitude of domain-specific data formats. Structured data is embedded into HTML pages using Microformats[7]. Various data providers have started to allow direct access to their databases via Web APIs.

So what is the rationale for adopting Linked Data instead of, or in addition to, these well-established publishing techniques? In summary, Linked Data provides a more generic, more flexible publishing paradigm which makes it easier for data consumers to discover and integrate data from large numbers of data sources. In particular, Linked Data provides:

- **A unifying data model**. Linked Data relies on RDF as a single, unifying data model. By providing for the globally unique identification of entities and by allowing different schemata

---

[7]http://microformats.org/

to be used in parallel to represent data, the RDF data model has been especially designed for the use case of global data sharing. In contrast, the other methods for publishing data on the Web rely on a wide variety of different data models, and the resulting heterogeneity needs to be bridged in the integration process.

- **A standardized data access mechanism.** Linked Data commits itself to a specific pattern of using the HTTP protocol. This agreement allows data sources to be accessed using generic data browsers and enables the complete data space to be crawled by search engines. In contrast, Web APIs are accessed using different proprietary interfaces.

- **Hyperlink-based data discovery**. By using URIs as global identifiers for entities, Linked Data allows hyperlinks to be set between entities in different data sources. These data links connect all Linked Data into a single global data space and enable Linked Data applications to discover new data sources at run-time. In contrast, Web APIs as well as data dumps in proprietary formats remain isolated data islands.

- **Self-descriptive data**. Linked Data eases the integration of data from different sources by relying on shared vocabularies, making the definitions of these vocabularies retrievable, and by allowing terms from different vocabularies to be connected to each other by vocabulary links.

Compared to the other methods of publishing data on the Web, these properties of the Linked Data architecture make it easier for data consumers to discover, access and integrate data. However, it is important to remember that the various publication methods represent a continuum of benefit, from making data available on the Web in any form, to publishing Linked Data according to the principles described in this chapter.

Progressive steps can be taken towards Linked Data publishing, each of which make it easier for third parties to consume and work with the data. These steps include making data available on the Web in any format but under an open license, to using structured, machine-readable formats that are preferably non-proprietary, to adoption of open standards such as RDF, and to inclusion of links to other data sources.

Tim Berners-Lee has described this continuum in terms of a five-star rating scheme [16], whereby data publishers can nominally award stars to their data sets according to the following criteria:

- 1 Star: data is available on the web (whatever format), but with an open license.

- 2 Stars: data is available as machine-readable structured data (e.g., Microsoft Excel instead of a scanned image of a table).

- 3 Stars: data is available as (2) but in a non-proprietary format (e.g., CSV instead of Excel).

- 4 Stars: data is available according to all the above, plus the use of open standards from the W3C (RDF and SPARQL) to identify things, so that people can link to it.

- 5 Stars: data is available according to all the above, plus outgoing links to other people's data to provide context.

Crucially, each rating can be obtained in turn, representing a progressive transition to Linked Data rather than a wholesale adoption in one operation.

.

CHAPTER 3

The Web of Data

A significant number of individuals and organisations have adopted Linked Data as a way to publish their data, not just placing it *on* the Web but using Linked Data to ground it *in* the Web [80]. The result is a global data space we call the *Web of Data* [30]. The Web of Data forms a *giant global graph* [17] consisting of billions of RDF statements from numerous sources covering all sorts of topics, such as geographic locations, people, companies, books, scientific publications, films, music, television and radio programmes, genes, proteins, drugs and clinical trials, statistical data, census results, online communities and reviews.

The Web of Data can be seen as an additional layer that is tightly interwoven with the classic document Web and has many of the same properties:

1. The Web of Data is generic and can contain any type of data.

2. Anyone can publish data to the Web of Data.

3. The Web of Data is able to represent disagreement and contradictionary information about an entity.

4. Entities are connected by RDF links, creating a global data graph that spans data sources and enables the discovery of new data sources. This means that applications do not have to be implemented against a fixed set of data sources, but they can discover new data sources at run-time by following RDF links.

5. Data publishers are not constrained in their choice of vocabularies with which to represent data.

6. Data is self-describing. If an application consuming Linked Data encounters data described with an unfamiliar vocabulary, the application can dereference the URIs that identify vocabulary terms in order to find their definition.

7. The use of HTTP as a standardized data access mechanism and RDF as a standardized data model simplifies data access compared to Web APIs, which rely on heterogeneous data models and access interfaces.

3.1   BOOTSTRAPPING THE WEB OF DATA

The origins of this Web of Data lie in the efforts of the Semantic Web research community and particularly in the activities of the W3C *Linking Open Data (LOD) project*[1], a grassroots community effort founded in January 2007. The founding aim of the project, which has spawned a vibrant and growing Linked Data community, was to bootstrap the Web of Data by identifying existing data sets available under open licenses, convert them to RDF according to the Linked Data principles, and to publish them on the Web. As a point of principle, the project has always been open to anyone who publishes data according to the Linked Data principles. This openness is a likely factor in the success of the project in bootstrapping the Web of Data.

Figure 3.1 and Figure 3.2 demonstrates how the number of data sets published on the Web as Linked Data has grown since the inception of the Linking Open Data project. Each node in the diagram represents a distinct data set published as Linked Data. The arcs indicate the existence of links between items in the two data sets. Heavier arcs correspond to a greater number of links, while bidirectional arcs indicate that outward links to the other exist in each data set.

Figure 3.2 illustrates the November 2010 scale of the *Linked Data Cloud* originating from the Linking Open Data project and classifies the data sets by topical domain, highlighting the diversity of data sets present in the Web of Data. The graphic shown in this figure is available online at `http://lod-cloud.net`. Updated versions of the graphic will be published on this website in regular intervals. More information about each of these data sets can be found by exploring the *LOD Cloud Data Catalog*[2] which is maintained by the LOD community within the *Comprehensive Knowledge Archive Network (CKAN)*[3], a generic catalog that lists open-license datasets represented using any format.

If you publish a linked data set yourself, please also add it to this catalog so that it will be included into the next version of the cloud diagram. Instructions on how to add data sets to the catalog are found in the ESW wiki[4].

3.2   TOPOLOGY OF THE WEB OF DATA

This section gives an overview of the topology of the Web of Data as of November 2010. Data sets are classified into the following topical domains: geographic, government, media, libraries, life science, retail and commerce, user-generated content, and cross-domain data sets. Table 3.1 gives an overview of the number of triples at this point in time, as well as the number of RDF links per domain. The number of RDF links refers to out-going links that are set from data sources within a domain to other data sources. The numbers are taken from the *State of the LOD Cloud* document[5]

---

[1]`http://esw.w3.org/topic/SweoIG/TaskForces/CommunityProjects/LinkingOpenData`
[2]`http://www.ckan.net/group/lodcloud`
[3]`http://www.ckan.net/`
[4]`http://esw.w3.org/TaskForces/CommunityProjects/LinkingOpenData/DataSets/CKANmetainformation`
[5]`http://lod-cloud.net/state/`

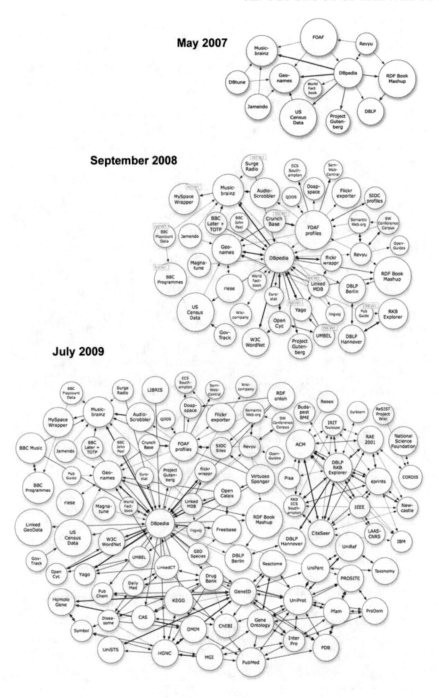

Figure 3.1: Growth in the number of data sets published on the Web as Linked Data.

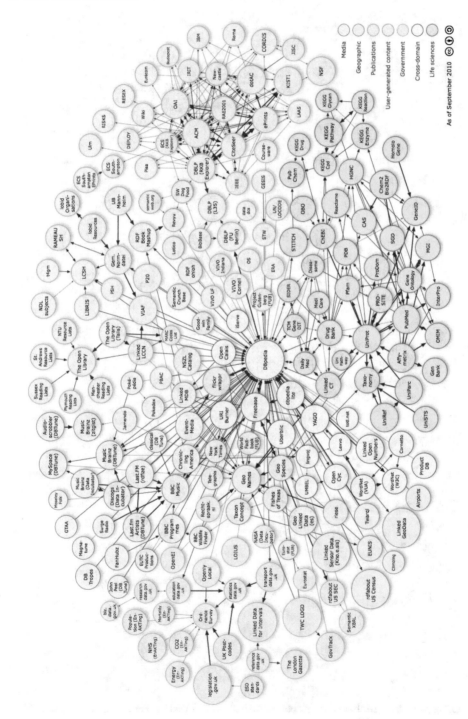

Figure 3.2: Linking Open Data cloud as of November 2010. The colors classify data sets by topical domain.

which on a regular basis compiles summary statistics about the data sets that are cataloged within the LOD Cloud Data Catalog on CKAN.

Domain	Data Sets	Triples	Percent	RDF Links	Percent
Cross-domain	20	1,999,085,950	7.42	29,105,638	7.36
Geographic	16	5,904,980,833	21.93	16,589,086	4.19
Government	25	11,613,525,437	43.12	17,658,869	4.46
Media	26	2,453,898,811	9.11	50,374,304	12.74
Libraries	67	2,237,435,732	8.31	77,951,898	19.71
Life sciences	42	2,664,119,184	9.89	200,417,873	50.67
User Content	7	57,463,756	0.21	3,402,228	0.86
	203	26,930,509,703		395,499,896	

3.2.1   CROSS-DOMAIN DATA

Some of the first data sets that appeared in the Web of Data are not specific to one topic, but span multiple domains. This cross-domain coverage is crucial for helping to connect domain-specific data sets into a single, interconnected data space, thereby avoiding fragmentation of the Web of Data into isolated, topical *data islands*. The prototypical example of cross-domain Linked Data is *DBpedia*[6] [32], a data set automatically extracted from publicly available Wikipedia dumps. Things that are the subject of a Wikipedia article are automatically assigned a DBpedia URI, based on the URI of that Wikipedia article. For example, the Wikipedia article about the city of Birmingham has the following URI `http://en.wikipedia.org/wiki/Birmingham`. Therefore, Birmingham has the corresponding DBpedia URI `http://dbpedia.org/resource/Birmingham` which is not the URI of a Web page about Birmingham , but a URI that identifies the city itself. RDF statements that refer to this URI are then generated by extracting information from various parts of the Wikipedia articles, in particular the *infoboxes* commonly seen on the right hand side of Wikipedia articles. Because of its breadth of topical coverage, DBpedia has served as a hub within the Web of Data from the early stages of the Linking Open Data project. The wealth of inward and outward links connecting items in DBpedia to items in other data sets is apparent in Figure 3.2.

A second major source of cross-domain Linked Data is *Freebase*[7], an editable, openly-licensed database populated through user contributions and data imports from sources such as Wikipedia and Geonames. Freebase provides RDF descriptions of items in the database, which are linked to items in DBpedia with incoming and outgoing links.

---

[6]`http://dbpedia.org/`
[7]`http://www.freebase.com`

Further cross-domain data sets include UMBEL[8], YAGO [104], and OpenCyc[9]. These are, in turn, linked with DBpedia, helping to facilitate data integration across a wide range of interlinked sources.

3.2.2   GEOGRAPHIC DATA

Geography is another factor that can often connect information from varied topical domains. This is apparent in the Web of Data, where the *Geonames*[10] data set frequently serves as a hub for other data sets that have some geographical component. Geonames is an open-license geographical database that publishes Linked Data about 8 million locations.

A second significant data set in this area is *LinkedGeoData* [102], a Linked Data conversion of data from the *OpenStreetMap* project, which provided information about more than 350 million spatial features. Wherever possible, locations in Geonames and LinkedGeoData are interlinked with corresponding locations in DBpedia, ensuring there is a core of interlinked data about geographical locations.

Linked Data versions of the EuroStat[11], World Factbook[12] and US Census[13] data sets begin to bridge the worlds of statistics, politics and social geography, while *Ordnance Survey* (the national mapping agency of Great Britain) has begun to publish Linked Data describing the administrative areas within the Great Britain[14], in efforts related to the *data.gov.uk* initiative described below.

3.2.3   MEDIA DATA

One of the first large organisations to recognise the potential of Linked Data and adopt the principles and technologies into their publishing and content management workflows has been the British Broadcasting Corporation (BBC). Following earlier experiments with publishing their catalogue of programmes as RDF, the BBC released in 2008 two large sites that combine publication of Linked Data and conventional Web pages. The first of these, */programmes*[15] provides a URI for and RDF description of every episode of every TV or radio programme broadcast across the BBC's various channels [71].

The second of these sites, */music*[16], publishes Linked Data about every artist whose music has been played on BBC radio stations, including incoming links from the specific programme episode during which it was broadcasted. This music data is interlinked with DBpedia, and it receives incoming links from a range of music-related Linked Data sources. These cross-data set links allow applications to consume data from all these sources and integrate it to provide rich artist profiles,

---

[8]http://umbel.org/
[9]http://sw.opencyc.org/
[10]http://www.geonames.org/
[11]http://ckan.net/package?q=eurostat&groups=lodcloud
[12]http://www4.wiwiss.fu-berlin.de/factbook/
[13]http://www.rdfabout.com/demo/census/
[14]http://data.ordnancesurvey.co.uk/
[15]http://www.bbc.co.uk/programmes
[16]http://www.bbc.co.uk/music

while the playlist data can be mined to find similarities between artists that may be used to generate recommendations.

More recently, the BBC have launched the site *Wildlife Finder*[17], which presents itself to users as a conventional Web site with extensive information about animal species, behaviours and habitats. Behind the scenes, each of these is identified by an HTTP URI and described in RDF. Outgoing links connect each species, behaviour and habitat to the corresponding resources in the DBpedia data set, and to BBC Programmes that depict these.

An indicator of the potential for Linked Data technologies within the enterprise, as well as on the public Web, comes from the BBC's World Cup 2010 Web site[18]. This high-traffic, public-facing Web site was populated with data modelled in RDF and stored in an RDF triple store[19]. In this case, the goal of using RDF was not to expose Linked Data for consumption by third parties, but to aid internal content management and data integration in a domain with high levels of connectivity between players, teams, fixtures and stadia.

Elsewhere in the media sector, there have also been significant moves towards Linked Data by major players. The New York Times has published a significant proportion of its internal subject headings as Linked Data[20] under a *Creative Commons Attribution* license (see Section 4.3.3), interlinking these topics with DBpedia, Freebase and Geonames. The intention is to use this liberally-licensed data as a map to lead people to the rich archive of content maintained by the New York Times.

As early as 2008, Thomson Reuters launched *Calais*[21], a Web service capable of annotating documents with the URIs of entities (e.g., places, people and companies) mentioned in the text. In many cases these Calais URIs are linked to equivalent URIs elsewhere in the Web of Data, such as to DBpedia and the CIA World Factbook. Services such as this are particularly significant for their ability to bridge Linked Data and conventional hypertext documents, potentially allowing documents such as blog posts or news articles to be enhanced with relevant pictures or background data.

3.2.4   GOVERNMENT DATA

Governmental bodies and public-sector organisations produce a wealth of data, ranging from economic statistics, to registers of companies and land ownership, reports on the performance of schools, crime statistics, and the voting records of elected representatives. Recent drives to increase government transparency, most notably in countries such as Australia[22], New Zealand[23], the U.K.[24] and

---

[17]http://www.bbc.co.uk/wildlifefinder/
[18]http://news.bbc.co.uk/sport1/hi/football/world_cup_2010/default.stm
[19]http://www.bbc.co.uk/blogs/bbcinternet/2010/07/bbc_world_cup_2010_dynamic_sem.html
[20]http://data.nytimes.com/
[21]http://www.opencalais.com/
[22]http://data.australia.gov.au/
[23]http://www.data.govt.nz/
[24]http://data.gov.uk

U.S.A.[25], have led to a significant increase in the amount of governmental and public-sector data that is made accessible on the Web. Making this data easily accessible enables organisations and members of the public to work with the data, analyse it to discover new insights, and build tools that help communicate these findings to others, thereby helping citizens make informed choices and hold public servants to account.

The potential of Linked Data for easing the access to government data is increasingly understood, with both the *data.gov.uk*[26] and *data.gov*[27] initiatives publishing significant volumes of data in RDF. The approach taken in the two countries differs slightly: to date the latter has converted very large volumes of data, while the former has focused on the creation of core data-level infrastructure for publishing Linked Data, such as stable URIs to which increasing amounts of data can be connected [101].

In a very interesting initiative is being pursued by the *UK Civil Service*[28] which has started to mark up job vacancies using RDFa. By providing information about open positions in a structured form, it becomes easier for external job portals to incorporate civil service jobs [25]. If more organizations would follow this example, the transparency in the labor market could be significantly increased [31].

Further high-level guidance on "Putting Government Data online" can be found in [18]. In order to provide a forum for coordinating the work on using Linked Data and other Web standards to improve access to government data and increase government transparency, W3C has formed a eGovernment Interest Group[29].

3.2.5   LIBRARIES AND EDUCATION

With an imperative to support novel means of discovery, and a wealth of experience in producing high-quality structured data, libraries are natural complementors to Linked Data. This field has seen some significant early developments which aim at integrating library catalogs on a global scale; interlinking the content of multiple library catalogs, for instance, by topic, location, or historical period; interlink library catalogs with third party information (picture and video archives, or knowledge bases like DBpedia); and at making library data easier accessible by relying on Web standards.

Examples include the American Library of Congress and the German National Library of Economics which publish their subject heading taxonomies as Linked Data (see [30] and [86], respectively), while the complete content of *LIBRIS* and the Swedish National Union Catalogue is available as Linked Data[31][32]. Similarly, the *OpenLibrary*, a collaborative effort to create "one Web

---

[25]http://www.data.gov
[26]http://data.gov.uk/linked-data
[27]http://www.data.gov/semantic
[28]http://www.civilservice.gov.uk/
[29]http://www.w3.org/egov/wiki/Main_Page
[30]http://id.loc.gov/authorities/about.html
[31]http://blog.libris.kb.se/semweb/?p=7
[32]http://blogs.talis.com/nodalities/2009/01/libris-linked-library-data.php

page for every book ever published"[33] publishes its catalogue in RDF, with incoming links from data sets such as *ProductDB* (see Section 3.2.7 below).

Scholarly articles from journals and conferences are also well represented in the Web of Data through community publishing efforts such as DBLP as Linked Data[34][35][36], RKBexplorer[37], and the Semantic Web Dogfood Server[38] [84].

An application that facilitates this scholarly data space is *Talis Aspire*[39]. The application supports educators in the creation and management of literature lists for university courses. Items are added to these lists through a conventional Web interface; however, behind the scenes, the system stores these records as RDF and makes the lists available as Linked Data. Aspire is used by various universities in the UK, which, in turn, have become Linked Data providers. The Aspire application is explored in more detail in Section 6.1.2.

High levels of ongoing activity in the library community will no doubt lead to further significant Linked Data deployments in this area. Of particular note in this area is the new *Object Reuse and Exchange (OAI-ORE)* standard from the *Open Archives Initiative* [110], which is based on the Linked Data principles. The OAI-ORE, Dublin Core, SKOS, and FOAF vocabularies form the foundation of the new Europeana Data Model[40]. The adoption of this model by libraries, museums and cultural institutions that participate in Europeana will further accelerate the availability of Linked Data related to publications and cultural heritage artifacts.

In order to provide a forum and to coordinate the efforts to increase the global interoperability of library data, W3C has started a Library Linked Data Incubator Group[41].

3.2.6   LIFE SCIENCES DATA

Linked Data has gained significant uptake in the Life Sciences as a technology to connect the various data sets that are used by researchers in this field. In particular, the Bio2RDF project [11] has interlinked more than 30 widely used data sets, including UniProt (the Universal Protein Resource), KEGG (the Kyoto Encyclopedia of Genes and Genomes), CAS (the Chemical Abstracts Service), PubMed, and the Gene Ontology. The W3C *Linking Open Drug Data* effort[42] has brought together the pharmaceutical companies Eli Lilly, AstraZeneca, and Johnson & Johnson, in a cooperative effort to interlink openly-licensed data about drugs and clinical trials, in order to aid drug-discovery [68].

---

[33]http://openlibrary.org/
[34]http://dblp.l3s.de/
[35]http://www4.wiwiss.fu-berlin.de/dblp/
[36]http://dblp.rkbexplorer.com/
[37]http://www.rkbexplorer.com/data/
[38]http://data.semanticweb.org/
[39]http://www.talis.com/aspire
[40]http://version1.europeana.eu/c/document_library/get_file?uuid=9783319c-9049-436c-bdf9-25f72e85e34c&groupId=10602
[41]http://www.w3.org/2005/Incubator/lld/
[42]http://esw.w3.org/HCLSIG/LODD

3.2.7   RETAIL AND COMMERCE

The *RDF Book Mashup*[43] [29] provided an early example of publishing Linked Data related to retail and commerce. The Book Mashup uses the Simple Commerce Vocabulary[44] to represent and republish data about book offers retrieved from the Amazon.com and Google Base Web APIs.

More recently, the *GoodRelations* ontology[45] [63] has provided a richer ontology for describing many aspects of e-commerce, such as businesses, products and services, offerings, opening hours, and prices. GoodRelations has seen significant uptake from retailers such as *Best Buy*[46] and *Overstock.com*[47] seeking to increase their visibility in search engines such as *Yahoo!* and *Google*, that recognise data published in RDFa using certain vocabularies and use this data to enhance search results (see Section 6.1.1.2). The adoption of the GoodRelations ontology has even extended to the publication of price lists for courses offered by *The Open University*[48].

The *ProductDB* Web site and data set[49] aggregates and links data about products for a range of different sources and demonstrates the potential of Linked Data for the area of product data integration.

3.2.8   USER GENERATED CONTENT AND SOCIAL MEDIA

Some of the earliest data sets in the Web of Data were based on conversions of, or wrappers around, *Web 2.0* sites with large volumes of *user-generated content*. This has produced data sets and services such as DBpedia and the *FlickrWrappr*[50], a Linked Data wrapper around the Flickr photo-sharing service. These were complemented by user-generated content sites that were built with native support for Linked Data, such as *Revyu.com* [61] for reviews and ratings, and Faviki[51] for annotating Web content with Linked Data URIs. Wiki systems that provide for publishing structured content as Linked Data on the Web include *Semantic MediaWiki*[52] and *Ontowiki*[53]. There are several hundred publicly accessible Semantic MediaWiki installations[54] that publish their content to the Web of Data.

More recently, Linked Data principles and technologies have been adopted by major players in the user-generated content and social media spheres, the most significant example of which is the development and adoption by *Facebook* of the *Open Graph Protocol*[55]. The Open Graph Protocol enables Web publishers to express a few basic pieces of information about the items described in

---

[43]http://www4.wiwiss.fu-berlin.de/bizer/bookmashup/
[44]http://sites.wiwiss.fu-berlin.de/suhl/bizer/bookmashup/simpleCommerceVocab01.rdf#
[45]http://purl.org/goodrelations/
[46]http://www.bestbuy.com/
[47]http://www.overstock.com/
[48]http://data.open.ac.uk/
[49]http://productdb.org/
[50]http://www4.wiwiss.fu-berlin.de/flickrwrappr/
[51]http://www.faviki.com/
[52]http://semantic-mediawiki.org/wiki/Semantic_MediaWiki
[53]http://ontowiki.net/Projects/OntoWiki
[54]http://semantic-mediawiki.org/wiki/Sites_using_Semantic_MediaWiki
[55]http://opengraphprotocol.org/

their Web pages, using RDFa (see Section 2.4.2). This enables Facebook to more easily consume data from sites across the Web, as it is published at source in structured form. Within a few months of its launch, numerous major destination sites on the Web, such as the *Internet Movie Database*[56], had adopted the Open Graph Protocol to publish structured data describing items featured on their Web pages. The primary challenge for the Open Graph Protocol is to enable a greater degree of linking between data sources, within the framework that has already been well established.

Another area in which RDFa is enabling the publication of user-generated content as Linked Data is through the *Drupal* content management system[57]. Version 7 of Drupal enables the description of Drupal *entities*, such as *users*, in RDFa.

3.3   CONCLUSIONS

The data sets described in this chapter demonstrate the diversity in the Web of Data. Recently published data sets, such as *Ordnance Survey, legislation.gov.uk*, the *BBC,* and the *New York Times* data sets, demonstrate how the Web of Data is evolving from data publication primarily by third party enthusiasts and researchers, to data publication *at source* by large media and public sector organisations. This trend is expected to gather significant momentum, with organisations in other industry sectors publishing their own data according to the Linked Data principles.

Linked Data is made available on the Web using a wide variety of tools and publishing patterns. In the following Chapters 4 and 5, we will examine the design decisions that must be taken to ensure your Linked Data sits well in the Web, and the technical options available for publishing it.

---

[56]http://www.imdb.com/
[57]http://drupal.org/

CHAPTER 4

Linked Data Design Considerations

So far this book has introduced the basic principles of Linked Data (Chapter 2) and given an overview of how these principles are being applied to the publication of data from a wide variety of domains (Chapter 3). This chapter will discuss the primary design considerations that must be taken into account when preparing data to be published as Linked Data on the Web, before introducing specific publishing recipes in Chapter 5.

These design considerations are not about visual design, but about how one shapes and structures data to fit neatly in the Web. They break down into three areas, each of which maps onto one or two of the Linked Data principles: (1) naming things with URIs; (2) describing things with RDF; (3) and making links to other data sets.

The outcome of these design decisions contributes directly to the utility and usability of a set of Linked Data, and therefore ultimately its value to the people and software programs that use it.

4.1 USING URIS AS NAMES FOR THINGS

As discussed in Chapter 2, the first principle of Linked Data is that URIs should be used as names for things that feature in your data set. These things might be concrete *real-world* entities such as a person, a building, your dog, or more *abstract* notions such as a scientific concept. Each of these things needs a name so that you and others can refer to it. Just as significant care should go into the design of URIs for pages in a conventional Web site, so should careful decisions be made about the design of URIs for a set of Linked Data. This section will explore these issues in detail.

4.1.1 MINTING HTTP URIS

The second principle of Linked Data is that URIs should be created using the `http://` URI scheme. This allows these *names* to be looked up using any client, such as a Web browser, that speaks the HTTP protocol.

In practical terms, using `http://` URIs as names for things simply amounts to a data publisher choosing part of an `http://` namespace that she controls, by virtue of owning the domain name, running a Web server at that location, and *minting* URIs in this *namespace* to identify the things in her data set.

For example, *Big Lynx* owns the domain name `biglynx.co.uk` and runs a Web server at `http://biglynx.co.uk/`. Therefore, they are free to *mint* URIs in this *namespace* to use as names for things they want to talk about. If *Big Lynx* wish to mint URIs to identify members of staff, they may do this in the namespace `http://biglynx.co.uk/people/`.

4.1.2   GUIDELINES FOR CREATING COOL URIS

As discussed in Chapter 1, a primary reason for publishing Linked Data is to add value through creation of incoming and outgoing links. Therefore, to help inspire confidence in third parties considering linking to a data set, some effort should be expended on minting stable, persistent URIs for entities in that data set. The specifics of the technical hosting environment may introduce some constraints on the precise syntax of these URIs; however, the following simple rules should be followed to help achieve this:

4.1.2.1   Keep out of namespaces you do not control

Where a particular Web site is seen as authoritative in a particular domain, and it provides stable URIs for entities in this domain (or pages about those entities), it can be very tempting to try and misappropriate these URIs for use in a Linked Data context. A common example of this is the *Internet Movie Database (IMDb)* [1], which has extensive data about films, actors, directors etc. Each is described in a document at an address such as:

- `http://www.imdb.com/title/tt0057012/`

which is the URI of a document about the film *Dr. Strangelove*.

It is not unreasonable at first glance to consider *augmenting* this URI with a fragment identifier to create a URI that identifies the film itself, rather than a document about the film, such as:

- `http://www.imdb.com/title/tt0057012/#film`

However, this approach is problematic as no one other than the owner of the `imdb.com` domain can make this URI dereferenceable, control what is returned when that URI is dereferenced, or ensure that it persists over time.

The recommended alternative in a Linked Data context is to mint one's own URI for each film, and state equivalence between these and corresponding URIs in other data sets where possible. If IMDb adopted the Linked Data principles it would constitute a highly appropriate target for such linking. However, this is not the case at the time of writing, and therefore alternatives such as DBpedia and *LinkedMDB* [2] should be considered.

4.1.2.2   Abstract away from implementation details

Wherever possible, URIs should not reflect implementation details that may need to change at some point in the future. For example, including server names or other indicators of underlying technical

---

[1] `http://www.imdb.com/`
[2] `http://linkedmdb.org/`

infrastructure in URIs is undesirable. In the case of *Big Lynx*, whose site is hosted on a machine called `tiger` and implemented mostly in PHP, the following is considered *uncool* as a potential URI for an RDF document containing data about Dave Smith, as it includes both the name of the machine and the `.php` extension:

- `http://tiger.biglynx.co.uk/people.php?id=dave-smith&format=rdf`

Similarly, publishers should avoid including port numbers in URIs, such as:

- `http://tiger.biglynx.co.uk:8080/people.php?id=dave-smith&format=rdf`

In contrast, the URI below could be considered cool, as it is less likely to break if the site is moved to a different machine or is reimplemented using a different scripting language or framework:

- `http://biglynx.co.uk/people/dave-smith.rdf`

The `mod_rewrite` module for the Apache Web server[3], and equivalents for other Web servers, allows you to configure your Web server such that implementation details are not exposed in URIs.

4.1.2.3 Use Natural Keys within URIs

To ensure the uniqueness of URIs it is often useful to base them on some existing primary key, such as a unique product ID in a database table. In the case of *Big Lynx*, the company is small enough that a combination of given name and family name can ensure uniqueness of URIs for members of staff, as shown in the examples below. This has the advantage of creating a more human-readable and memorable URI. In a larger organisation, an employee ID number may provide a suitable alternative.

A good general principle is to, wherever possible, use a key that is meaningful within the domain of the data set. For the sake of example (and if we ignore for the moment the issues with non-uniqueness of ISBNs), using the ISBN as part of the URI for a book is preferable to using its primary key from your internal database. This makes it significantly easier to link your book data with that from other sources as there is a common key on which links can be based. Linking bibliographic works, including the use of natural versus articifical keys, is discussed in more detail in [103].

References [15] and [98] provide background information on the topic of minting Cool URIs and are recommended reading.

4.1.3 EXAMPLE URIS

Each entity represented in a particular data set will likely lead to the minting of at least three URIs, as discussed in Section 2.3.1:

1. a URI for the real-world object itself

2. a URI for a related information resource that describes the real-world object and has an HTML representation

---

[3]`http://httpd.apache.org/docs/2.0/mod/mod_rewrite.html`

3. a URI for a related information resource that describes the real-world object and has an RDF/XML representation

Common syntactic forms for these URIs include the following examples from DBpedia:

1. `http://dbpedia.org/resource/Wildlife_photography`

2. `http://dbpedia.org/page/Wildlife_photography`

3. `http://dbpedia.org/data/Wildlife_photography`

It has one major disadvantage in that the resource URI is not very visually distinct from those of the RDF and HTML representations of the associated description. This can be problematic for developers new to Linked Data concepts, as they may not realise that the URI in a browser address bar has changed following content negotiation and a 303 redirect, and inadvertently refer to the wrong URI.

An alternative form is:

1. `http://id.biglynx.co.uk/dave-smith`

2. `http://pages.biglynx.co.uk/dave-smith`

3. `http://data.biglynx.co.uk/dave-smith`

This form has the advantage that the various URIs are more visually distinct due to the use of different subdomains. From an system architectural perspective, this may also simplify the Linked Data publication process by allowing RDF descriptions of resources to be served by a D2R Server (described in Section 5.2.4) on one machine at the `data` subdomain, while custom scripts on another machine at the `pages` subdomain could render sophisticated HTML documents describing the resources. Scripts at the `id` subdomain would simply be responsible for performing content negotiation and 303 redirects.

A third URI pattern that is also regularly used by Linked Data sources is to distinguish document URIs from concept URIs by adding the respective file extensions at the end of the URIs.

1. `http://biglynx.co.uk/people/dave-smith`

2. `http://biglynx.co.uk/people/dave-smith.html`

3. `http://biglynx.co.uk/people/dave-smith.rdf`

4.2   DESCRIBING THINGS WITH RDF

The third principle of Linked Data states "When someone looks up a URI, provide useful information...". Assuming that a URI has been minted for each entity in a data set, according to the guidelines above, the next consideration concerns what information to provide in response when someone looks up that URI.

Let us assume that we have a data set expressed as RDF triples. Which of these triples should be included in the RDF description of a particular resource? The list below enumerates the various types of triples that should be included into the description:

1. Triples that describe the resource with literals

2. Triples that describe the resource by linking to other resources (e.g., triples stating the resource's creator, or its type)

3. Triples that describe the resource by linking from other resources (i.e., incoming links)

4. Triples describing related resources (e.g., the name and maybe affiliation of the resource's creator)

5. Triples describing the description itself (i.e., data about the data, such as its provenance, date of collection, or licensing terms)

6. Triples about the broader data set of which this description is a part.

4.2.1 LITERAL TRIPLES AND OUTGOING LINKS

The significance of items 1 and 2 in the list above is self apparent; they constitute all the triples within your data set which have the resource's URI as the subject, and therefore together provide a basic description of the resource.

The sample code from a staff profile for Dave Smith, first shown in Section 2.5.1 and extended here, demonstrates describing a resource with literals and outgoing links.

```
1   @prefix rdf: <http://www.w3.org/1999/02/22-rdf-syntax-ns#> .
2   @prefix foaf: <http://xmlns.com/foaf/0.1/> .
3   @prefix rel: <http://purl.org/vocab/relationship/> .
4
5   <http://biglynx.co.uk/people/dave-smith>
6       rdf:type foaf:Person ;
7       foaf:name "Dave Smith" ;
8       foaf:based_near <http://sws.geonames.org/3333125/> ;
9       foaf:based_near <http://dbpedia.org/resource/Birmingham> ;
10      foaf:topic_interest <http://dbpedia.org/resource/Wildlife_photography> ;
11      foaf:knows <http://dbpedia.org/resource/David_Attenborough> ;
12      rel:employerOf <http://biglynx.co.uk/people/matt-briggs> .
```

Of particular importance are triples that provide human-readable labels for resources that can be used within client applications. Predicates such as `rdfs:label`, `foaf:name`, `skos:prefLabel` and `dcterms:title` should be used for this purpose as they are widely supported by Linked Data applications. In cases where a comment or textual description of the resource is available, these should be published using predicates such as `dcterms:description` or `rdfs:comment`.

4.2.2 INCOMING LINKS

If an RDF triple links person *a* to person *b*, the document describing *b* should include this triple, which can be thought of as an *incoming link* to *b* (item 3 on the list above). This helps ensure that data about person *a* is discoverable from the description of *b*, even if *a* is not the object of any triples in the description of *b*. For instance, when you use a Linked Data browser to navigate from resource *a* to *b*, incoming links enable you to navigate backward to resource *a*. They also enable crawlers of Linked Data search engines, which have entered a Linked Data site via an external Link pointing at resource *b*, to discover resource *a* and continue crawling the site.

The code sample below demonstrates this principle applied to the *Big Lynx* scenario. In this case the code shows both an outgoing *employerOf* link from Dave Smith to Matt Briggs and the inverse *employedBy* link from Matt Briggs to Dave Smith. Following this principle, the same two triples would be published in the document describing Matt Briggs.

```
1   @prefix rdf: <http://www.w3.org/1999/02/22-rdf-syntax-ns#> .
2   @prefix foaf: <http://xmlns.com/foaf/0.1/> .
3   @prefix rel: <http://purl.org/vocab/relationship/> .
4
5   <http://biglynx.co.uk/people/dave-smith>
6     rdf:type foaf:Person ;
7     foaf:name "Dave Smith" ;
8     foaf:based_near <http://sws.geonames.org/3333125/> ;
9     foaf:based_near <http://dbpedia.org/resource/Birmingham> ;
10    foaf:topic_interest <http://dbpedia.org/resource/Wildlife_photography> ;
11    foaf:knows <http://dbpedia.org/resource/David_Attenborough> ;
12    rel:employerOf <http://biglynx.co.uk/people/matt-briggs> ;
13    foaf:isPrimaryTopicOf <http://biglynx.co.uk/people/dave-smith.rdf> .
14
15  <http://biglynx.co.uk/people/dave-smith.rdf>
16    foaf:primaryTopic <http://biglynx.co.uk/people/dave-smith> .
17
18  <http://biglynx.co.uk/people/matt-briggs>
19    rel:employedBy <http://biglynx.co.uk/people/dave-smith> .
```

In addition, a `foaf:primaryTopic` and a `foaf:isPrimaryTopicOf` triple has been added to the example. These triples connect the URI of the RDF document describing Dave to the URI that is used to identify him as a real-world object, and thus make the relationship between the two explicit [98].

It should be noted that while including incoming links in the description of a resource is deemed good practice, there may be occasions where this is unfeasible due to the number of incoming links. In such cases the data publisher must exercise judgement in publishing those incoming links that may be most useful to data consumers.

4.2.3 TRIPLES THAT DESCRIBE RELATED RESOURCES

In addition to including triples in an RDF document that link the resource described by that document to related resources within the same data set, it may be desirable to include some additional triples describing these related resources. For example, an RDF document describing a *Big Lynx* employee, as described above, could usefully link to the organisational team of which that individual is a member, and provide an RDF triple giving the name of that team. An extension of this principle could be to include links to the other team members and a small amount of information about each of them.

Various forms of *Concise Bounded Description* have been defined[4] that characterise the extent and nature of related triples that should be included in the description of a resource. This issue (item 4 on the above list) is slightly controversial. One school of thought says that each RDF description of a resource should convey the same informational content as its corresponding HTML page or contain enough data such that applications consuming the RDF description do not also have to request the descriptions of all resources referenced in the description before they can begin displaying information to the user. For example, an application interested in information about one employee may well be interested in information about others in the same team. Providing this information at the earliest opportunity may save the application from making additional HTTP requests for related documents, with the associated latency, before they can display a coherent set of information to the user.

The opposing view argues that replicating data about resource B (or even resource C) in the description of resource A introduces redundant data that consuming applications must process in order to reconcile or discard. This may be trivial where the provenance of information is of relatively little importance, but where any computationally-intensive process is applied to de-duplication or ranking of information, this may serve as a significant disadvantage. In addition, while certain basic data, such as triples giving the *label* or *type* of a resource, may be required by every application consuming the data, the data publisher cannot anticipate every potential use of the data and therefore cannot easily make meaningful decisions regarding what additional data about B should be included in a description of A.

As is often the case, there are no hard and fast rules regarding this issue, and each data publisher should consider the specifics of their publication scenario and the potential consumption patterns of the data.

4.2.4   TRIPLES THAT DESCRIBE THE DESCRIPTION

In contrast to item 4, there is nothing controversial about items 5 and 6 (triples describing the description itself, and the parent data set). Unfortunately this guidance is often overlooked, reducing the discoverability of related data and preventing data consumers from gaining background information about the data set, such as its licensing terms. We will explore these issues in detail in the next section.

---

[4]http://n2.talis.com/wiki/Bounded_Descriptions_in_RDF

4.3   PUBLISHING DATA ABOUT DATA

4.3.1   DESCRIBING A DATA SET

In addition to ensuring that resources featured in a data set are richly described, the same principles should be applied to the data set itself, to include information about authorship of a data set, its currency (i.e., how recently the data set was updated), and its licensing terms. This *metadata* gives consumers of the data clarity about the provenance and timeliness of a data set and the terms under which it can be reused; each, of which, are important in encouraging reuse of a data set.

Furthermore, descriptions of a data set can include pointers to example resources within the data set, thereby aiding discovery and indexing of the data by crawlers. If the load created by crawlers indexing a site is too great, descriptions of a data set can also include links to RDF data dumps, which can be downloaded and indexed separately.

Two primary mechanisms are available for publishing descriptions of a data set: *Semantic Sitemaps* [43] and *voiD* [5] descriptions.

4.3.1.1   Semantic Sitemaps

Semantic Sitemaps [5] are an extension to the well-established *Sitemaps* protocol [6] that provides search engines with hints about pages in a Web site that are available for crawling. A Sitemap consists of an XML document, which is typically named `sitemap.xml` and stored in the root directory of a Web site. The Sitemaps schema defines elements such as `url`, `loc`, `lastmod`, and `changefreq` that enable a site owner to convey basic information about pages that make up a site and the rate at which they typically change, thereby allowing search engines to optimise their crawling behaviour.

The Semantic Sitemaps extension defines additional elements that can be used to enhance a Sitemap XML file with information relevant in a Linked Data context, such as a label and URI for the data set, sample URIs that feature in the data set, plus the location of corresponding SPARQL endpoints and data dumps. By using the `datasetURI` element in a Sitemap, data publishers can inform search engines and other client applications of the URI of the data set itself, from where they may be able to retrieve additional descriptive information in RDF.

The code below shows an example Semantic Sitemap for the *Big Lynx* Web site that explicitly references a data set describing people who work at *Big Lynx*, in addition to providing generic information such as the location of data dumps and the company's main SPARQL endpoint:

```
1  <?xml version="1.0" encoding="UTF-8"?>
2  <urlset
3    xmlns=
4     "http://www.sitemaps.org/schemas/sitemap/0.9"
5    xmlns:sc=
6     "http://sw.deri.org/2007/07/sitemapextension/scschema.xsd"
7  >
8    <sc:dataset>
9      <sc:datasetLabel>Big Lynx People Data Set</sc:datasetLabel>
```

[5]http://sw.deri.org/2007/07/sitemapextension/
[6]http://www.sitemaps.org/

```
10      <sc:datasetURI>
11         http://biglynx.co.uk/datasets/people
12      </sc:datasetURI>
13      <sc:linkedDataPrefix>
14         http://biglynx.co.uk/people/
15      </sc:linkedDataPrefix>
16      <sc:sampleURI>
17         http://biglynx.co.uk/people/dave-smith
18      </sc:sampleURI>
19      <sc:sampleURI>
20         http://biglynx.co.uk/people/matt-briggs
21      </sc:sampleURI>
22      <sc:sparqlEndpointLocation>
23         http://biglynx.co.uk/sparql
24      </sc:sparqlEndpointLocation>
25      <sc:dataDumpLocation>
26         http://biglynx.co.uk/dumps/people.rdf.gz
27      </sc:dataDumpLocation>
28      <changefreq>monthly</changefreq>
29    </sc:dataset>
30  </urlset>
```

If the data publisher wishes to convey to a consumer the shape of the graph that can be expected when a URI is dereferenced (for example, a *Concise Bounded Description* or a *Symmetric Concise Bounded Description*), the optional `slicing` attribute can be added to the `sc:linkedDataPrefix`.

```
1  <sc:linkedDataPrefix slicing="cbd">
2     http://biglynx.co.uk/people/
3  </sc:linkedDataPrefix>
```

Acceptable values for this attribute include `cbd` and `scbd`. A full list of these values, with explanatory notes, can be found at [7].

4.3.1.2 voiD Descriptions

voiD (the Vocabulary of Interlinked Datasets)[8] is the de facto standard vocabulary for describing Linked Data sets. It replicates some of the capabilities of Semantic Sitemaps (e.g., the terms `dataDump`, `sparqlEndpoint`) but does so in RDF. voiD also enables the vocabularies used in a data set and the links it has to others to be described, as well as logical partitions (or *subsets*) of a specific data set.

The ability to define and link subsets of data is particularly useful, as it allows rich RDF data to be published about individual RDF descriptions, which can in turn be defined as subsets of a broader data set. In such cases, the subject of RDF triples should be the URI of the RDF description itself (i.e., an information resource), not the URI of the resource it describes.

For example, Nelly Jones may be the creator of Dave Smith's online profile in RDF (at `http://biglynx.co.uk/people/dave-smith.rdf`), but she is not the creator of Dave Smith himself. Therefore, the URI `http://biglynx.co.uk/people/dave-smith.rdf` should be used as the subject of

---

[7]http://sw.deri.org/2007/07/sitemapextension/#slicing
[8]http://www.w3.org/2001/sw/interest/void/

triples about this RDF description, rather than the URI identifying Dave Smith. This is illustrated in the following example:

```
1   @prefix rdf: <http://www.w3.org/1999/02/22-rdf-syntax-ns#> .
2   @prefix rdfs: <http://www.w3.org/2000/01/rdf-schema#> .
3   @prefix dcterms: <http://purl.org/dc/terms/> .
4   @prefix foaf: <http://xmlns.com/foaf/0.1/> .
5   @prefix rel: <http://purl.org/vocab/relationship/> .
6
7   <http://biglynx.co.uk/people/dave-smith.rdf>
8     foaf:primaryTopic <http://biglynx.co.uk/people/dave-smith> ;
9     rdf:type foaf:PersonalProfileDocument ;
10    rdfs:label "Dave Smith's Personal Profile in RDF" ;
11    dcterms:creator <http://biglynx.co.uk/people/nelly-jones> ;
12
13  <http://biglynx.co.uk/people/dave-smith>
14    foaf:isPrimaryTopicOf <http://biglynx.co.uk/people/dave-smith.rdf> ;
15    rdf:type foaf:Person ;
16    foaf:name "Dave Smith" ;
17    foaf:based_near <http://sws.geonames.org/3333125/> ;
18    foaf:based_near <http://dbpedia.org/resource/Birmingham> ;
19    foaf:topic_interest <http://dbpedia.org/resource/Wildlife_photography> ;
20    foaf:knows <http://dbpedia.org/resource/David_Attenborough> ;
21    rel:employerOf <http://biglynx.co.uk/people/matt-briggs> .
22
23  <http://biglynx.co.uk/people/matt-briggs>
24        rel:employedBy <http://biglynx.co.uk/people/dave-smith> .
```

Note that these triples could be included in a document published anywhere on the Web. However, to ensure the online profile of Dave Smith is as self-describing as possible, they should appear within the RDF document itself, at http://biglynx.co.uk/people/dave-smith.rdf. It should also be noted that, given convention regarding the use of the .rdf extension, the online profile should be published using the RDF/XML serialisation of RDF. However, for the sake of readability this and subsequent examples are shown using the more-readable Turtle serialisation.

This example is extended below to include the publisher of the staff profile document (in this case *Big Lynx*, identified by the URI http://biglynx.co.uk/company.rdf#company) and its date of publication. A triple has also been added stating that Dave Smith's staff profile document *is part of* the broader People data set introduced in Section 4.3.1.1 above, and it is identified by the URI http://biglynx.co.uk/datasets/people:

```
1   @prefix rdf: <http://www.w3.org/1999/02/22-rdf-syntax-ns#> .
2   @prefix rdfs: <http://www.w3.org/2000/01/rdf-schema#> .
3   @prefix xsd: <http://www.w3.org/2001/XMLSchema#> .
4   @prefix dcterms: <http://purl.org/dc/terms/> .
5   @prefix foaf: <http://xmlns.com/foaf/0.1/> .
6   @prefix rel: <http://purl.org/vocab/relationship/> .
7   @prefix void: <http://rdfs.org/ns/void#> .
8
9   <http://biglynx.co.uk/people/dave-smith.rdf>
10    foaf:primaryTopic <http://biglynx.co.uk/people/dave-smith> ;
11    rdf:type foaf:PersonalProfileDocument ;
12    rdfs:label "Dave Smith's Personal Profile in RDF" ;
```

```
13      dcterms:creator  <http://biglynx.co.uk/people/nelly-jones>  ;
14      dcterms:publisher  <http://biglynx.co.uk/company.rdf#company>  ;
15      dcterms:date  "2010-11-05"^^xsd:date  ;
16      dcterms:isPartOf  <http://biglynx.co.uk/datasets/people>  .
17
18   <http://biglynx.co.uk/people/dave-smith>
19      foaf:isPrimaryTopicOf  <http://biglynx.co.uk/people/dave-smith.rdf>  ;
20      rdf:type  foaf:Person  ;
21      foaf:name  "Dave Smith"  ;
22      foaf:based_near  <http://sws.geonames.org/3333125/>  ;
23      foaf:based_near  <http://dbpedia.org/resource/Birmingham>  ;
24      foaf:topic_interest  <http://dbpedia.org/resource/Wildlife_photography>  ;
25      foaf:knows  <http://dbpedia.org/resource/David_Attenborough>  ;
26      rel:employerOf  <http://biglynx.co.uk/people/matt-briggs>  .
27
28   <http://biglynx.co.uk/people/matt-briggs>
29      rel:employedBy  <http://biglynx.co.uk/people/dave-smith>  .
```

Having linked the staff profile document at http://biglynx.co.uk/people/dave-smith.rdf to the broader People data set http://biglynx.co.uk/datasets/people, a corresponding description should be provided of that data set. Following convention, this will be published at http://biglynx.co.uk/datasets/people.rdf, and may contain the triples shown in the example below:

```
1    @prefix rdf: <http://www.w3.org/1999/02/22-rdf-syntax-ns#> .
2    @prefix rdfs: <http://www.w3.org/2000/01/rdf-schema#> .
3    @prefix dcterms: <http://purl.org/dc/terms/> .
4    @prefix foaf: <http://xmlns.com/foaf/0.1/> .
5    @prefix void: <http://rdfs.org/ns/void#> .
6
7    <http://biglynx.co.uk/datasets/people.rdf>
8       foaf:primaryTopic  <http://biglynx.co.uk/datasets/people>  ;
9       rdf:type  foaf:Document  ;
10      rdfs:label  "Description of the Big Lynx People Data Set"  .
11
12   <http://biglynx.co.uk/datasets/people>
13      foaf:isPrimaryTopicOf  <http://biglynx.co.uk/datasets/people.rdf>  ;
14      rdf:type  void:Dataset  ;
15      dcterms:title  "Big Lynx People Data Set"  ;
16      rdfs:label  "Big Lynx People Data Set"  ;
17      dcterms:description  "Data set describing people who work at Big Lynx"  ;
18      void:exampleResource  <http://biglynx.co.uk/people/dave-smith>  ;
19      void:exampleResource  <http://biglynx.co.uk/people/matt-briggs>  ;
20      void:exampleResource  <http://biglynx.co.uk/people/nelly-jones>  ;
21      dcterms:hasPart  <http://biglynx.co.uk/people/dave-smith.rdf>  ;
22      dcterms:isPartOf  <http://biglynx.co.uk/datasets/all>  .
23
24   <http://biglynx.co.uk/datasets/master>
25      void:subset  <http://biglynx.co.uk/datasets/people>  .
```

The People data set is itself a subset of the entire *Big Lynx* Master data set, which is briefly mentioned at the bottom of the example above, to demonstrate use of the void:subset property to create an incoming link. Note the directionality of this property, which points from a super data set to one of its subsets, not the other way around. At present, no inverse of `void:subset` has been defined, although a term such as `void:inDataset` has been discussed to address this issue. In the meantime, the term `dcterms:isPartOf` makes a reasonable substitute, and therefore is used in the examples above.

4.3.2   PROVENANCE METADATA

The ability to track the origin of data is a key component in building trustworthy, reliable applications [38]. The use of dereferenceable URIs hard-wires this capability into Linked Data, as anyone can dereference a particular URI to determine what the owner of that namespace says about a particular resource. However, as different information providers might publish data within the same namespace, it is important to be able to track the origin of particular data fragments. Therefore, data sources should publish provenance meta data together with the data itself. Such meta data should be represented as RDF triples describing the document in which the original data is contained.

A widely deployed vocabulary for representing such data is Dublin Core[9], particularly the `dc:creator`, `dc:publisher` and `dc:date` predicates. When using the `dc:creator`, `dc:publisher` properties in the Linked Data context, you should use the URIs and not the literal names identifying the creator and publisher. This allows others to unambiguously refer to them and, for instance, connect these URIs with background information about them which is available on the Web and might be used to assess the quality and trustworthiness of published data. Dave Smith's personal profile shown above demonstrates the use of these predicates for conveying simple provenance data.

The Open Provenance Model[10] provides an alternative, more expressive vocabulary, that describes provenance in terms of *Agents*, *Artifacts* and *Processes*. A comparison of different provenance vocabularies as well as further resources regarding publication of provenance information are available on the website of the *W3C Provenance Incubator Group*[11].

In order to enable data consumers to verify attribution metadata, publishers may decide to digitally sign their data. An open-source library that can be used to produce such signatures is the NG4J - Named Graphs API for Jena[12].

4.3.3   LICENSES, WAIVERS AND NORMS FOR DATA

It is a common assumption that *content* (e.g., blog posts, photos) and data made publicly available on the Web can be reused at will. However, the absence of a licensing statement does not grant consumers the automatic right to use that content/data. Conversely, it is relatively common for Web publishers to omit licensing statements from data published online due to an explicit desire to share that data.

Unfortunately, the absence of clarity for data consumers about the terms under which they can reuse a particular data set is likely to hinder reuse of that data, as limited investment will be

---

[9]http://dublincore.org/documents/dcmi-terms/
[10]http://purl.org/net/opmv/ns
[11]http://www.w3.org/2005/Incubator/prov/
[12]http://www4.wiwiss.fu-berlin.de/bizer/ng4j/

made in building applications over data whose terms of reuse are unclear. Therefore, all Linked Data published on the Web should include explicit *license* or *waiver* statements [82].

Up to this point, a consumer of the data in Dave Smith's staff profile would be aware of its creator and date of publication, as this is expressed in the data itself. The data consumer would not, however, have any information about the terms under which they could reuse the data in their own applications. For example, a third-party software developer may wish to aggregate such profiles and build a directory of people working in the television production industry but could not be certain that the publisher of Dave Smith's profile consents to such usage and if any conditions are attached. To address this issue, it is highly important that additional triples are added to RDF documents describing the license or waiver under which the data they contain is made available.

4.3.3.1 Licenses vs. Waivers

Licenses and waivers represent two sides of the same coin: licenses grant others rights to reuse something and generally attach conditions to this reuse, while waivers enable the owner to explicitly waive their rights to something, such as a data set. Before selecting a license or waiver to apply to a specific data set, the data publisher must understand the types of licenses or waivers that are legally suited to that data set.

Perhaps the licenses in most widespread usage on the Web are those developed by the *Creative Commons* initiative[13], which allow content owners to attach conditions, such as *attribution*, to the reuse of their work. The legal basis for the Creative Commons licenses is *copyright*, which is applicable to *creative works*. Precisely defining a creative work, from a legal perspective, is beyond the scope of this book. However, by way of example, a photo or a blog post could be considered a creative work, while factual information such as the geo-coordinates of the *Big Lynx* headquarters would not, thereby making a Creative Commons license inapplicable.

To put it another way, one cannot copyright facts. Consequently, any license based on copyright, such as the Creative Commons licenses, is inapplicable to factual data. While one could apply such a license to indicate intent regarding how the data is used, it would have no legal basis.

The following sections illustrate the application of licenses and waivers to copyrightable and non-copyrightable material, respectively.

4.3.3.2 Applying Licenses to Copyrightable Material

The *Big Lynx* Web site includes a *blog* where staff members publish updates about the company and accounts of production expeditions. For example, the *Big Lynx* Lead Cameraman, Matt Briggs, used the blog to share his experiences of filming for the series *Pacific Sharks*, in a post with the URI `http://biglynx.co.uk/blog/making-pacific-sharks`. As with all *Big Lynx* blog posts, this will be published under the "Attribution-Share Alike 3.0 Unported" license (often referred to by the shorthand *CC-BY-SA*), which requires those who reuse the content to attribute the creator when they reuse it and apply the same license to any derivative works.

---

[13]`http://creativecommons.org/`

The code below shows an extract of the Creative Commons CC-BY-SA license expressed in RDF. The RDF/XML code at[14] has been converted to the Turtle serialisation of RDF for readability.

```
 1   @prefix rdf: <http://www.w3.org/1999/02/22-rdf-syntax-ns#> .
 2   @prefix dc: <http://purl.org/dc/elements/1.1/> .
 3   @prefix cc: <http://creativecommons.org/ns#> .
 4
 5   <http://creativecommons.org/licenses/by-sa/3.0/>
 6     rdf:type cc:License ;
 7     dc:title "Attribution-Share Alike 3.0 Unported" ;
 8     dc:creator <http://creativecommons.org/> ;
 9     cc:permits
10       cc:Distribution ,
11       cc:DerivativeWorks ,
12       cc:Reproduction ;
13     cc:requires
14       cc:ShareAlike ,
15       cc:Attribution ,
16       cc:Notice ;
17
18     xhv:alternate
19       <http://creativecommons.org/licenses/by-sa/3.0/deed.af> ,
20       ...
21       <http://creativecommons.org/licenses/by-sa/3.0/rdf> .
```

Applying this license to content published on the Web and described in RDF is simple, involving the addition of just one RDF triple. The code sample below shows an RDF description of the "Making Pacific Sharks" blog post. The post is described using the SIOC ontology [36] of online communities. Note that a URI, `http://biglynx.co.uk/blog/making-pacific-sharks`, has been minted to identify the post itself, which is reproduced and described in corresponding RDF and HTML documents.

The `cc:license` triple in the code associates the CC-BY-SA license with the blog post, using the vocabulary that the Creative Commons has defined for this purpose[15].

```
 1   @prefix rdf: <http://www.w3.org/1999/02/22-rdf-syntax-ns#> .
 2   @prefix rdfs: <http://www.w3.org/2000/01/rdf-schema#> .
 3   @prefix dcterms: <http://purl.org/dc/terms/> .
 4   @prefix foaf: <http://xmlns.com/foaf/0.1/> .
 5   @prefix sioc: <http://rdfs.org/sioc/ns#> .
 6   @prefix cc: <http://creativecommons.org/ns#> .
 7
 8   <http://biglynx.co.uk/blog/making-pacific-sharks>
 9     rdf:type sioc:Post ;
10     dcterms:title "Making Pacific Sharks" ;
11     dcterms:date "2010-11-10T16:34:15Z"^^xsd:dateTime ;
12     sioc:has_container <http://biglynx.co.uk/blog/> ;
13     sioc:has_creator <http://biglynx.co.uk/people/matt-briggs> ;
14     foaf:primaryTopic <http://biglynx.co.uk/productions/pacific-sharks> ;
15     sioc:topic <http://biglynx.co.uk/productions/pacific-sharks> ;
16     sioc:topic <http://biglynx.co.uk/locations/pacific-ocean> ;
```

[14] http://creativecommons.org/licenses/by-sa/3.0/rdf
[15] http://creativecommons.org/ns#

```
17    sioc:topic <http://biglynx.co.uk/species/shark> ;
18    sioc:content "Day one of the expedition was a shocker – monumental swell,
         tropical storms, and not a shark in sight. The Pacific was up to its old
         tricks again. I wasn't holding out hope of filming in the following 48
         hours, when the unexpected happened..." ;
19    foaf:isPrimaryTopicOf <http://biglynx.co.uk/blog/making-pacific-sharks.rdf> ;
20    foaf:isPrimaryTopicOf <http://biglynx.co.uk/blog/making-pacific-sharks.html> ;
21    cc:license <http://creativecommons.org/licenses/by-sa/3.0/> .
22
23   <http://biglynx.co.uk/blog/making-pacific-sharks.rdf>
24     rdf:type foaf:Document ;
25     dcterms:title "Making Pacific Sharks (RDF version)" ;
26     foaf:primaryTopic <http://biglynx.co.uk/blog/making-pacific-sharks> .
27
28   <http://biglynx.co.uk/blog/making-pacific-sharks.html>
29     rdf:type foaf:Document ;
30     dcterms:title "Making Pacific Sharks (HTML version)" ;
31     foaf:primaryTopic <http://biglynx.co.uk/blog/making-pacific-sharks> .
```

Note how the license has been applied to the blog post itself, not the documents in which it is reproduced. This is because the documents in which the post is reproduced contain both copyrightable and non-copyrightable material (e.g., the post content and the date of publication, respectively) and therefore cannot be covered meaningfully by the license applied to the blog post itself.

4.3.3.3 Non-copyrightable Material

The procedure for applying a waiver to non-copyrightable material is similar to that shown above for copyrightable material. The two primary differences are: 1. the predicate used to link the material with the waiver – here we use the dedicated *Waiver Vocabulary*[16]; 2. the addition of the norms predicate.

Norms provide a means for data publishers who waive their legal rights (through application of a waiver) to define *expectations* they have about how the data is used. For example, a data publisher may waive their rights to a data set yet still wish to be attributed as the source of the data in cases where it is reused and republished.

The code sample below shows the application of the *Open Data Commons Public Domain Dedication and Licence* as a waiver to Dave Smith's staff profile document, extending the sample used in Section 4.3.1. The *Open Data Commons Attribution-Share Alike* norm (ODC-BY-SA) is also applied, using the norms predicate from the Waiver Vocabulary. This signifies the data publisher's desire for attribution when the data is reused and that the same norms are applied to derivative works. There are obvious parallels between this norm and the CC-BY-SA license, with the key difference being that CC-BY-SA sets legal requirements while ODC-BY-SA sets social expectations.

```
1    @prefix rdf: <http://www.w3.org/1999/02/22-rdf-syntax-ns#> .
2    @prefix rdfs: <http://www.w3.org/2000/01/rdf-schema#> .
3    @prefix xsd: <http://www.w3.org/2001/XMLSchema#> .
```

[16]http://vocab.org/waiver/terms/

```
 4   @prefix dcterms: <http://purl.org/dc/terms/> .
 5   @prefix foaf: <http://xmlns.com/foaf/0.1/> .
 6   @prefix rel: <http://purl.org/vocab/relationship/> .
 7   @prefix void: <http://rdfs.org/ns/void#> .
 8   @prefix wv: <http://vocab.org/waiver/terms/> .
 9
10   <http://biglynx.co.uk/people/dave-smith.rdf>
11     foaf:primaryTopic <http://biglynx.co.uk/people/dave-smith> ;
12     rdf:type foaf:PersonalProfileDocument ;
13     rdfs:label "Dave Smith's Personal Profile in RDF" ;
14     dcterms:creator <http://biglynx.co.uk/people/nelly-jones> ;
15     dcterms:publisher <http://biglynx.co.uk/company.rdf#company> ;
16     dcterms:date "2010-11-05"^^xsd:date ;
17     dcterms:isPartOf <http://biglynx.co.uk/datasets/people> ;
18     wv:waiver
19       <http://www.opendatacommons.org/odc-public-domain-dedication-and-licence/> ;
20     wv:norms
21       <http://www.opendatacommons.org/norms/odc-by-sa/> .
22
23   <http://biglynx.co.uk/people/dave-smith>
24     foaf:isPrimaryTopicOf <http://biglynx.co.uk/people/dave-smith.rdf> ;
25     rdf:type foaf:Person ;
26     foaf:name "Dave Smith" ;
27     foaf:based_near <http://sws.geonames.org/3333125/> ;
28     foaf:based_near <http://dbpedia.org/resource/Birmingham> ;
29     foaf:topic_interest <http://dbpedia.org/resource/Wildlife_photography> ;
30     foaf:knows <http://dbpedia.org/resource/David_Attenborough> ;
31     rel:employerOf <http://biglynx.co.uk/people/matt-briggs> .
32
33   <http://biglynx.co.uk/people/matt-briggs>
34     rel:employedBy <http://biglynx.co.uk/people/dave-smith> .
```

Note that in cases where an entire data set contains purely non-copyrightable material it would also be prudent to explicitly apply the waiver to the entire data set, in addition to the individual RDF documents that make up that data set.

4.4   CHOOSING AND USING VOCABULARIES TO DESCRIBE DATA

RDF provides a generic, abstract data model for describing resources using *subject, predicate, object* triples. However, it does not provide any *domain-specific terms* for describing classes of things in the world and how they relate to each other. This function is served by *taxonomies, vocabularies* and *ontologies* expressed in *SKOS* (Simple Knowledge Organization System) [81], *RDFS* (the RDF Vocabulary Description Language, also known as RDF Schema) [37] and *OWL* (the Web Ontology Language) [79].

4.4.1   SKOS, RDFS AND OWL

SKOS is a vocabulary for expressing conceptual hierarchies, often referred to as taxonomies, while RDFS and OWL provide vocabularies for describing conceptual models in terms of classes and their properties. For example, someone may define an RDFS vocabulary about pets that includes a class Dog, of which all individual dogs are members. They may also define a property hasColour, thereby allowing people to publish RDF descriptions of their own dogs using these terms.

Collectively, SKOS, RDFS and OWL provide a continuum of expressivity. SKOS is widely used to represent thesauri, taxonomies, subject heading systems, and topical hierarchies (for instance that mechanics *belong to* the boarder topic of physics). RDFS and OWL are used in cases where subsumption relationships between terms should be represented (for instance that all athletes *are also* persons). When paired with a suitable reasoning engine, RDFS and OWL models allow implicit relationships to be inferred for the data. In a Linked Data context, it is often sufficient to express vocabularies in RDFS. However, certain primitives from OWL, such as sameAs, are used regularly to state that two URIs identify the same resource, as discussed in 4.5. The combination of RDFS plus certain OWL primitives is often referred to colloquially as *RDFS++*.

A full discussion of SKOS, RDFS and OWL is beyond the scope of this book – for that the reader should refer to Allemang and Hendler's highly recommended book *Semantic Web for the Working Ontologist* [6]. However, a brief overview of the basics of RDFS is important as a foundation for further discussions in the remainder of this book.

4.4.2   RDFS BASICS

RDFS is a language for describing lightweight ontologies in RDF; these are often referred to as *vocabularies*. In their most simple form, RDFS vocabularies consist of *class* and *property* type definitions, such as in the example above of the class *Dog* (of which there may be many instances) and the property *hasColour*.

For historic reasons, the primitives of the RDFS language are defined in two separate namespaces:

- The http://www.w3.org/2000/01/rdf-schema# namespace is associated (by convention) with the rdfs: namespace prefix

- The http://www.w3.org/1999/02/22-rdf-syntax-ns# namespace is associated (by convention) with the rdf: namespace prefix

The two basic classes within the RDFS language are:

- *rdfs:Class* which is the class of resources that are RDF classes

- *rdf:Property* which is the class of all RDF properties

An RDF resource is declared to be a class by typing it as an instance of rdfs:Class using the rdf:type predicate. The code below shows a simple RDFS vocabulary for describing TV productions.

This vocabulary duplicates some aspects of the well-established *Programmes Ontology*, developed at the BBC[17]. The Programmes Ontology would be a better choice for publishing data in a

---

[17]http://purl.org/ontology/po/

non-fictional scenario, for the reasons discussed in Section 4.4.4 on *Reusing Existing Terms*. However, the example below is provided for simplicity of illustration. Namespace prefixes are used to improve the readability of the code that defines the vocabulary. It should be noted, however, that the classes and properties are themselves resources, whose URIs should be made dereferenceable according to the same Linked Data principles that apply to data described using the vocabulary [23].

```
1   @prefix rdf: <http://www.w3.org/1999/02/22-rdf-syntax-ns#> .
2   @prefix rdfs: <http://www.w3.org/2000/01/rdf-schema#> .
3   @prefix owl: <http://www.w3.org/2002/07/owl#> .
4   @prefix dcterms: <http://purl.org/dc/terms/> .
5   @prefix foaf: <http://xmlns.com/foaf/0.1/> .
6   @prefix cc: <http://creativecommons.org/ns#> .
7   @prefix prod: <http://biglynx.co.uk/vocab/productions#> .
8
9   <>
10    rdf:type owl:Ontology ;
11    rdfs:label "Big Lynx Productions Vocabulary" ;
12    dcterms:creator <http://biglynx.co.uk/people/nelly-jones> ;
13    dcterms:publisher <http://biglynx.co.uk/company.rdf#company> ;
14    dcterms:date "2010-10-31"^^xsd:date ;
15    cc:license <http://creativecommons.org/licenses/by-sa/3.0/> .
16
17  prod:Production
18    rdf:type rdfs:Class ;
19    rdfs:label "Production";
20    rdfs:comment "the class of all productions".
21
22  prod:Director
23    rdf:type rdfs:Class ;
24    rdfs:label "Director";
25    rdfs:comment "the class of all directors";
26    rdfs:subClassOf foaf:Person .
27
28  prod:director
29    rdf:type owl:ObjectProperty ;
30    rdfs:label "director";
31    rdfs:comment "the director of the production";
32    rdfs:domain prod:Production ;
33    rdfs:range prod:Director ;
34    owl:inverseOf prod:directed .
35
36  prod:directed
37    rdf:type owl:ObjectProperty ;
38    rdfs:label "directed";
39    rdfs:comment "the production that has been directed";
40    rdfs:domain prod:Director ;
41    rdfs:range prod:Production ;
42    rdfs:subPropertyOf foaf:made .
```

The terms `prod:Production` and `prod:Director` are declared to be classes by typing them as instances of `rdfs:Class`. The term `prod:director` is declared to be a property by typing it as an instance of `rdf:Property`.

The staff profile of Matt Briggs, the Lead Cameraman at *Big Lynx*, shows how these terms can be used to describe Matt Briggs's role as a director (Line 17) and, specifically, as director of *Pacific Sharks* (Line 21):

```
1   @prefix rdf: <http://www.w3.org/1999/02/22-rdf-syntax-ns#> .
2   @prefix rdfs: <http://www.w3.org/2000/01/rdf-schema#> .
3   @prefix dcterms: <http://purl.org/dc/terms/> .
4   @prefix foaf: <http://xmlns.com/foaf/0.1/> .
5   @prefix prod: <http://biglynx.co.uk/vocab/productions#> .
6
7   <http://biglynx.co.uk/people/matt-briggs.rdf>
8     foaf:primaryTopic <http://biglynx.co.uk/people/matt-briggs> ;
9     rdf:type foaf:PersonalProfileDocument ;
10    rdfs:label "Matt Brigg's Personal Profile in RDF" ;
11    dcterms:creator <http://biglynx.co.uk/people/nelly-jones> ;
12
13  <http://biglynx.co.uk/people/matt-briggs>
14    foaf:isPrimaryTopicOf <http://biglynx.co.uk/people/matt-briggs.rdf> ;
15    rdf:type
16      foaf:Person ,
17      prod:Director ;
18    foaf:name "Matt Briggs" ;
19    foaf:based_near <http://sws.geonames.org/3333125/> ;
20    foaf:based_near <http://dbpedia.org/resource/Birmingham> ;
21    foaf:topic_interest <http://dbpedia.org/resource/Wildlife_photography> ;
22    prod:directed <http://biglynx.co.uk/productions/pacific-sharks> .
```

4.4.2.1   Annotations in RDFS

RDFS defines two properties for annotating resources:

- `rdfs:label` may be used to provide a human-readable name for a resource.

- `rdfs:comment` may be used to provide a human-readable description of a resource.

Use of both of these properties when defining RDFS vocabularies is recommended, as they provide guidance to potential users of the vocabulary and are relied upon by many Linked Data applications when displaying data. In addition to being used to annotate terms in RDFS vocabularies, these properties are also commonly used to provide labels and descriptions for other types of RDF resources.

4.4.2.2   Relating Classes and Properties

RDFS also provides primitives for describing relationships between classes and between properties.

- `rdfs:subClassOf` is used to state that all the instances of one class are also instances of another. In the example above, `prod:Director` is declared to be a subclass of the `Person` class from the *Friend of a Friend (FOAF)* ontology. This has the implication that all instances of the class `prod:Director` are also instances of the class `foaf:Person`.

- `rdfs:subPropertyOf` is used to state that resources related by one property are also related by another. In the example vocabulary above, the property `prod:directed` is a subproperty of `foaf:made`, meaning that a director who directed a production also *made* that production.

- `rdfs:domain` is used to state that *any resource that has a given property* is an instance of one or more classes. The domain of the `prod:director` property defined above is declared as `prod:Production`, meaning that all resources which are described using the `prod:director` property are instances of the class `prod:Production`.

- `rdfs:range` is used to state that *all values of a property* are instances of one or more classes. In the example above, the range of the `prod:director` property is declared as `prod:Director`. Therefore, a triple stating `<a> prod:director <b>` implies that `<b>` is an instance of the class `prod:Director`.

By using these relational primitives, the author of an RDFS vocabulary implicitly defines rules that allow additional information to be inferred from RDF graphs. For instance, the rule that all *directors* are also *people*, enables the triple `<http://biglynx.co.uk/people/matt-briggs> rdf:type foaf:Person` to be inferred from the triple `<http://biglynx.co.uk/people/matt-briggs> rdf:type prod:Director`. The result is that not all relations need to be created explicitly in the original data set, as many can be inferred based on axioms in the vocabulary. This can simplify the management of data in Linked Data applications without compromising the comprehensiveness of a data set.

4.4.3   A LITTLE OWL

OWL extends the expressivity of RDFS with additional modeling primitives. For example, OWL defines the primitives `owl:equivalentClass` and `owl:equivalentProperty`. When combined with `rdfs:subClassOf` and `rdfs:subPropertyOf`, these provide powerful mechanisms for defining mappings between terms from different vocabularies, which, in turn, increase the interoperability of data sets modeled using different vocabularies, as described in Section 2.5.3.

Another OWL modeling primitive that is very useful in the context of the Web of Data is `owl:InverseFunctionalProperty`. The values of properties that are declared to be inverse-functional uniquely identify the thing having the property. Thus, by declaring properties, such as `foaf:icqChatID` or `foaf:openid`, to be inverse functional properties, vocabulary maintainers can help Linked Data applications to perform identity resolution (see Section 6.3.3).

The property `owl:inverseOf` allows the creator of a vocabulary to state that one property is the inverse of another, as demonstrated in the example above where it is stated that `prod:directed` is the `owl:inverseOf` `tv:director`. In practical terms this means that for every triple of the form `<production> prod:director <director>`, a triple of the form `<director> tv:directed <production>` can be inferred. `owl:inverseOf` is also *symmetric*, meaning that if the property `prod:director` is the inverse of `prod:directed`, then `prod:directed` is also the inverse of `prod:director`.

4.4.4   REUSING EXISTING TERMS

If suitable terms can be found in existing vocabularies, these should be reused to describe data wherever possible, rather than reinvented. Reuse of existing terms is highly desirable as it maximises the probability that data can be consumed by applications that may be *tuned* to well-known vocabularies, without requiring further pre-processing of the data or modification of the application.

The following list presents a number of vocabularies that cover common types of data, are in widespread usage, and should be reused wherever possible.

- The **Dublin Core Metadata Initiative (DCMI) Metadata Terms** vocabulary[18] defines general metadata attributes such as *title*, *creator*, *date* and *subject*.

- The **Friend-of-a-Friend (FOAF)** vocabulary[19] defines terms for describing persons, their activities and their relations to other people and objects.

- The **Semantically-Interlinked Online Communities (SIOC)** vocabulary[20] (pronounced *"shock"*) is designed for describing aspects of online community sites, such as users, posts and forums.

- The **Description of a Project (DOAP)** vocabulary[21] (pronounced *"dope"*) defines terms for describing software projects, particularly those that are Open Source.

- The **Music Ontology**[22] defines terms for describing various aspects related to music, such as artists, albums, tracks, performances and arrangements.

- The **Programmes Ontology**[23] defines terms for describing programmes such as TV and radio broadcasts.

- The **Good Relations Ontology**[24] defines terms for describing products, services and other aspects relevant to e-commerce applications.

- The **Creative Commons (CC)** schema[25] defines terms for describing copyright licenses in RDF.

- The **Bibliographic Ontology (BIBO)**[26] provides concepts and properties for describing citations and bibliographic references (i.e., quotes, books, articles, etc.).

[18]http://dublincore.org/documents/dcmi-terms/
[19]http://xmlns.com/foaf/spec/
[20]http://rdfs.org/sioc/spec/
[21]http://trac.usefulinc.com/doap
[22]http://musicontology.com/
[23]http://www.bbc.co.uk/ontologies/programmes/2009-09-07.shtml
[24]http://purl.org/goodrelations/
[25]http://creativecommons.org/ns#
[26]http://bibliontology.com/

- The **OAI Object Reuse and Exchange** vocabulary[27] is used by various library and publication data sources to represent resource aggregations such as different editions of a document or its internal structure.

- The **Review Vocabulary**[28] provides a vocabulary for representing reviews and ratings, as are often applied to products and services.

- The **Basic Geo (WGS84)** vocabulary[29] defines terms such as *lat* and *long* for describing geographically-located things.

There will always be cases where new terms need to be developed to describe aspects of a particular data set [22], in which case these terms should be mapped to related terms in well-established vocabularies, as discussed in 2.5.3.

Where newly defined terms are specialisations of existing terms, there is an argument for using both terms in tandem when publishing data. For example, in the *Big Lynx* scenario, Nelly may decide to explicitly add RDF triples to Matt Briggs's profile stating that he is a foaf:Person as well as a prod:Director, even though this could be inferred by a reasoner based on the relationships defined in the *Big Lynx* Productions vocabulary. This practice can be seen as an instance of the *Materialize Inferences* pattern[30], and while it introduces an element of redundancy it also conveys the benefits described above whereby the accessibility of data to Linked Data applications that do not employ reasoning engines is maximised.

4.4.5   SELECTING VOCABULARIES

At the time of writing, there is no definitive directory that can be consulted to find suitable vocabularies and ontologies. However, *SchemaWeb*[31], *SchemaCache*[32], and *Swoogle*[33] provide useful starting points. Further insights into patterns and levels of vocabulary usage *in the wild* can be gained from the vocabulary usage statistics provided in Section 2.3 of the *State of the LOD Cloud* document[34].

In selecting vocabularies for reuse the following criteria should be applied:

1. **Usage and uptake** – is the vocabulary in widespread usage? Will using this vocabulary make a data set more or less accessible to existing Linked Data applications?

2. **Maintenance and governance** – is the vocabulary actively maintained according to a clear governance process? When, and on what basis, are updates made?

---

[27] http://www.openarchives.org/ore/
[28] http://purl.org/stuff/rev#
[29] http://www.w3.org/2003/01/geo/
[30] http://patterns.dataincubator.org/book/ch04s07.html
[31] http://www.schemaweb.info/
[32] http://schemacache.com/
[33] http://swoogle.umbc.edu/
[34] http://lod-cloud.net/state#terms

3. **Coverage** – does the vocabulary cover enough of the data set to justify adopting its terms and *ontological commitments*[35]?

4. **Expressivity** – is the degree of expressivity in the vocabulary appropriate to the data set and application scenario? Is it too expressive, or not expressive enough?

4.4.6   DEFINING TERMS

In cases where existing vocabularies are not adequate to describe a particular data set, new terms will need to be developed in a dedicated vocabulary, applying the features of RDFS and OWL outlined briefly in Section 4.4.2 above, and covered in significantly greater detail in [6]. The following aspects of best practice should be taken into consideration when defining vocabularies:

1. Supplement existing vocabularies rather than reinventing their terms.

2. Only define new terms in a namespace that you control.

3. Use terms from RDFS and OWL to relate new terms to those in existing vocabularies.

4. Apply the Linked Data principles equally rigorously to vocabularies as to data sets – URIs of terms should be dereferenceable so that Linked Data applications can look up their definition [23].

5. Document each new term with human-friendly labels and comments – `rdfs:label` and `rdfs:comment` are designed for this purpose.

6. Only define things that matter – for example, defining domains and ranges helps clarify how properties should be used, but over-specifying a vocabulary can also produce unexpected inferences when the data is consumed. Thus you should not overload vocabularies with ontological axioms, but better define terms rather loosely (for instance, by using only the RDFS and OWL terms introduced above).

A number of tools are available to assist with the vocabulary development process:

- **Neologism**[36] is a Web-based tool for creating, managing and publishing simple RDFS vocabularies. It is open-source and implemented in PHP on top of the Drupal-platform.

- **TopBraid Composer**[37] is a powerful commercial modeling environment for developing Semantic Web ontologies.

- **Protégé**[38] is an open-source ontology editor with a dedicated OWL plugin.

---

[35]http://en.wikipedia.org/wiki/Ontological_commitment
[36]http://neologism.deri.ie/
[37]http://www.topquadrant.com/products/TB_Composer.html
[38]http://protege.stanford.edu/

- The **NeOn Toolkit**[39] is an open-source ontology engineering environment with an extensive set of plugins.

4.5 MAKING LINKS WITH RDF

4.5.1 MAKING LINKS WITHIN A DATA SET

This section will explore the process of creating links within and between data sets. Both aspects are essential in ensuring that a data set is integrated with the Web at large, and that all resources it describes are fully discoverable once the data set has been located.

4.5.1.1 Publishing Incoming and Outgoing Links

Aside from small, static data sets that may reasonably be published on the Web in one RDF file, most data sets will be split across multiple RDF documents for publication as Linked Data. Whether these multiple documents are static or dynamically generated is less important than the structural characteristic whereby fragments of data from the same data set are spread across multiple documents on the same Web server.

In this situation it must be ensured that related resources (and the documents that describe them) are linked to each other, ensuring that each fragment of data may be discovered by crawlers or other applications consuming the data set through link traversal. This can be considered analogous to ensuring that a conventional Web site has adequate mechanisms for navigation between pages, such that there are no orphan pages in a site, and should be implemented according to the considerations described in Section 4.2.

4.5.2 MAKING LINKS WITH EXTERNAL DATA SOURCES

After publication of a set of Linked Data, it should be ensured that RDF links from external sources point at URIs in the data set. This helps to ensure that data can be discovered by RDF browsers and crawlers, and can be achieved by supplementing existing data sets (owned by the same publisher or by third parties) with RDF links pointing at resources in the new data set. Assuming that the existing data sets also have incoming links, then the new data set will be discoverable.

Third parties may need convincing of the value of linking to a new data set. Factors that may be persuasive in such situations are the value of the new data set (i.e., *is this data that was not previously available?*), the value it adds to the existing data set if linked (i.e., *what can be achieved that would not be possible without the new data set?*), and the cost of creating high quality links (i.e., *how complex is the creation and maintenance of such links?*).

One strategy is to create the necessary RDF links and ask third parties to include these triples in their data sets. This approach has been used successfully with DBpedia, which integrates various link sets generated by third parties that link DBpedia resources to those in other data sets[40].

[39]http://neon-toolkit.org/
[40]http://wiki.dbpedia.org/Downloads351#h120-1

4.5.2.1 Choosing External Linking Targets

Of equal importance to incoming links are outgoing links. Chapter 3 describes some of the wide variety of data sets already available on the Web of Data, which collectively provide many potential targets for links from within a new data set. The two main benefits of using URIs from these data sources are:

1. The URIs are dereferenceable, meaning that a description of the concept can be retrieved from the Web. For instance, using the DBpedia URI `http://dbpedia.org/resource/Birmingham` to identify the city of Birmingham provides an extensive description of the city, including abstracts in many languages.

2. The URIs are already linked to URIs from other data sources. For example, it is possible to navigate from the DBpedia URI `http://dbpedia.org/resource/Birmingham` to data about Birmingham provided by Geonames. Therefore, by linking to URIs from these data sets, data becomes connected into a rich and fast-growing network of other data sources.

A comprehensive list of data sets that may be suitable as linking targets is maintained in the CKAN repository[41]. In evaluating data sets as potential linking targets, it is important to consider the following questions [13]:

- What is the value of the data in the target data set?

- To what extent does this add value to the new data set?

- Is the target data set and its namespace under stable ownership and active maintenance?

- Are the URIs in the data set stable and unlikely to change?

- Are there ongoing links to other data set so that applications can tap into a network of interconnected data sources?

Once suitable linking targets have been identified, links can be created using the methods described in the remainder of this chapter.

4.5.2.2 Choosing Predicates for Linking

The nature of the data being published will determine which terms make suitable predicates for linking to other data sets. For instance, commonly used terms for linking in the domain of people are `foaf:knows`, `foaf:based_near` and `foaf:topic_interest`. Examples of combining these properties with property values from DBpedia, the DBLP bibliography and the RDF Book Mashup can be found in the online profiles of Tim Berners-Lee[42] and Ivan Herman[43].

In general, the factors that should be taken into account when choosing predicates for linking are:

[41]`http://ckan.net/group/lodcloud`
[42]`http://www.w3.org/People/Berners-Lee/card`
[43]`http://www.ivan-herman.net/foaf.rdf`

1. How widely is the predicate already used for linking by other data sources?

2. Is the vocabulary well maintained and properly published with dereferenceable URIs?

A list of widely used vocabularies from which linking properties can be chosen is given in Section 4.4.4 as well as in Section 2.4 of the *State of the LOD Cloud* document[44]. If very specific or proprietary terms are used for linking, they should be linked to more generic terms using `rdfs:subPropertyOf` mappings, as described in 2.5.3 and 4.4.4 as this enables client applications to translate them to a recognised vocabulary.

4.5.3   SETTING RDF LINKS MANUALLY

RDF links can be set manually or automatically – the choice of method will depend on the data set and the context in which it is published. Manual interlinking is typically employed for small, static data sets, while larger data sets generally require an automated or semi-automated approach.

Once target data sets have been identified, these can be manually searched to find the URIs of target resources for linking. If a data source doesn't provide a search interface, such as a SPARQL endpoint or a HTML Web form, a Linked Data browser can be used to explore the data set and find the relevant URIs.

Services such as Sindice[45] and Falcons[46] provide an index of URIs that can be searched by keyword and used to identify candidate URIs for linking. If multiple candidate URIs from different data sets are found, then links can be created to each of them, if they meet the criteria as linking targets. Alternatively, just one target data set may be chosen based on the criteria described in Section 4.5.2.1. Again, decisions such as these need to be taken based on the specifics of the publishing context.

It is important to remember that data sources use different URIs to identify real-world objects and the HTML or RDF documents describing these objects. A common mistake when setting links manually is to point at the document URIs and not at the URIs identifying the real-world object. Therefore, care should be taken when selecting target URIs to avoid unintentionally stating that a person lives in, or is friends, with a document.

4.5.4   AUTO-GENERATING RDF LINKS

The approach described above does not scale to larger data sets, for instance, interlinking 413,000 places in DBpedia to their corresponding entries in Geonames. The usual approach in such cases is to use automatic or semi-automatic record linkage heuristics to generate RDF links between the data sources. *Record linkage*, also called *identity resolution* or *duplicate detection*, is a well-known problem in databases [46] as well as in the ontology matching community [49], and many of the techniques from these fields are directly applicable in a Linked Data context.

---

[44] http://www4.wiwiss.fu-berlin.de/lodcloud/state/#terms
[45] http://sindice.com/
[46] http://iws.seu.edu.cn/services/falcons/objectsearch

In principle, there are two main types of record linkage techniques: simple key-based approaches that exploit common naming schemata used by both data sources; more complex, similarity-based approaches which compare data items and interlink them if their similarity is above a given threshold.

4.5.4.1 Key-based Approaches

In various domains, there are generally accepted naming schemata. For instance, *Global Trade Item Numbers (GTIN)* are commonly used to identify products; in the publication domain, there are ISBN numbers, in the financial domain there are ISIN identifiers. If a data set contains such identifiers, these should be exposed either as part of the URIs or as property values. Such properties are called *inverse functional properties* as their value uniquely identifies the subject of the triple and should be defined as such in the corresponding vocabulary, by stating they are of type `owl:InverseFunctionalProperty`.

Including commonly accepted identifiers in URIs, or as inverse functional properties into published data, lays the foundation for using simple pattern-based algorithms to generate RDF links between data. An example data source using GTIN codes for products in its URIs is *ProductDB*, which assigns the URI `http://productdb.org/gtin/09781853267802` to a particular version of *The Origin of Species* by Charles Darwin. URI aliases are also created based on the ISBN[47] and EAN[48] identifiers for the book, aiding further key-based linking.

4.5.4.2 Similarity-based Approaches

In cases where no common identifiers exist across data sets, it is necessary to employ more complex similarity-based linkage heuristics. These heuristics may compare multiple properties of the entities that are to be interlinked as well as properties of related entities. They aggregate the different similarity scores and interlink entities if the aggregated similarity value is above a given threshold. For instance, Geonames and DBpedia both provide information about geographic places. In order to identify places that appear in both data sets, one could use a heuristic that compares the names of places using a string similarity function, longitude and latitude values using a geographic matcher, the name of the country in which the places are located, as well as their population count. If all (or most) of the comparisons result in high similarity scores, it is assumed that both places are the same.

As one can not assume Web data sources provide complete descriptions of resources, similarity heuristics need to be chosen that tolerate missing values. DBpedia, for instance, only contains population counts for a fraction of the described places. An appropriate matching heuristic could therefore be to give additional weight to the country in which a place is located in cases where the population count is missing.

There are several tools available that allow matching heuristics to be defined in a declarative fashion and automate the process of generating RDF links based on these declarations.

---

[47]`http://productdb.org/isbn/9781853267802`
[48]`http://productdb.org/ean/9781853267802`

- **Silk - Link Discovery Framework** [111]. Silk provides a flexible, declarative language for specifying mathing heuristics. Mathing heuristics may combine different string matchers, numeric as well as geographic matchers. Silk enables data values to be transformed before they are used in the matching process and allows similarity scores to be aggregated using various aggregation functions. Silk can match local as well as remote datasets which are accessed using the SPARQL protocol. Matching tasks that require a large number of comparisons can be handled either by using different blocking features or by running Silk on a Hadoop cluster. Silk is available under the Apache License and can be downloaded from the project website[49]
.

- **LIMES - Link Discovery Framework for Metric Spaces** [44]. LIMES implements a fast and lossless approach for large-scale link discovery based on the characteristics of metric spaces but provides a less expressive language for specifying matching heuristics. Detailed information about LIMES is found on the project website[50].

In addition to the tools above, which rely on users explicitly specifying the matching heuristic, there are also tools available which learn the matching heuristic directly from the data. Examples of such tools include RiMOM[51], idMash [77], and ObjectCoref[52].

The advantage of learning matching heuristics is that the systems do not need to be manually configured for each type of links that are to be created between datasets. The disadvantage is that machine learning-based approaches typically have a lower precision compared to approaches that rely on domain knowledge provided by humans in the form of a matching description. The Instance Matching Track within Ontology Alignment Evaluation Initiative 2010[53] compared the quality of links that were produced by different learning based-tools. The evaluation revealed precision values between 0.6 and 0.97 and showed that quality of the resulting links depends highly on the specific linking task.

A task related to link generation is the maintenance of links over time as data sources change. There are various proposals for notification mechanisms to handle this task, an overview of which is given in [109]. In [87], the authors propose *DSNotify*, a framework that monitors Linked Data sources and informs consuming applications about changes. More information about Link Discovery tools and an up-to-date list of references is maintained by the LOD community at[54].

---

[49]http://www4.wiwiss.fu-berlin.de/bizer/silk/
[50]http://aksw.org/Projects/limes
[51]http://keg.cs.tsinghua.edu.cn/project/RiMOM/
[52]http://ws.nju.edu.cn/services/ObjectCoref
[53]http://www.dit.unitn.it/~p2p/OM-2010/oaei10_paper0.pdf
[54]http://esw.w3.org/TaskForces/CommunityProjects/LinkingOpenData/EquivalenceMining

CHAPTER 5

Recipes for Publishing Linked Data

This chapter will examine various common patterns for publishing Linked Data, which demonstrate how Linked Data complements rather than replaces existing data management infrastructures. Following this conceptual overview, the chapter introduces a series of recipes for publishing Linked Data on the Web that build on the design considerations outlined in 4 and use the *Big Lynx* scenario to illustrate the various approaches. The chapter concludes with tips for testing and debugging Linked Data published on the Web.

5.1 LINKED DATA PUBLISHING PATTERNS

Publishing Linked Data requires adoption of the basic principles outlined in Chapter 2. Compliance with the standards and best practices that underpin these principles is what enables Linked Data to streamline data interoperability and reuse over the Web. However, compliance with the Linked Data principles does not entail abandonment of existing data management systems and business applications but simply the addition of extra technical layer of glue to connect these into the Web of Data. While there is a very large number of technical systems that can be connected into the Web of Data, the mechanisms for doing so fall into a small number of Linked Data *publishing patterns*. In this section, we will give an overview of these patterns.

Figure 5.1 shows the most common Linked Data publishing patterns in the form of workflows, from structured data or textual content through to Linked Data published on the Web. In the following section, we will briefly address some of the key features of the workflows in 5.1.

5.1.1 PATTERNS IN A NUTSHELL

The primary consideration in selecting a workflow for publishing Linked Data concerns the nature of the input data.

5.1.1.1 From Queryable Structured Data to Linked Data

Data sets stored in relational databases can be published relatively easily as Linked Data through the use of relational database to RDF wrappers. These tools allow the data publisher to define mappings from relational database structures to RDF graphs that are served up on the Web according to the Linked Data principles. Section 5.2.4 gives an overview of relational database to RDF wrappers.

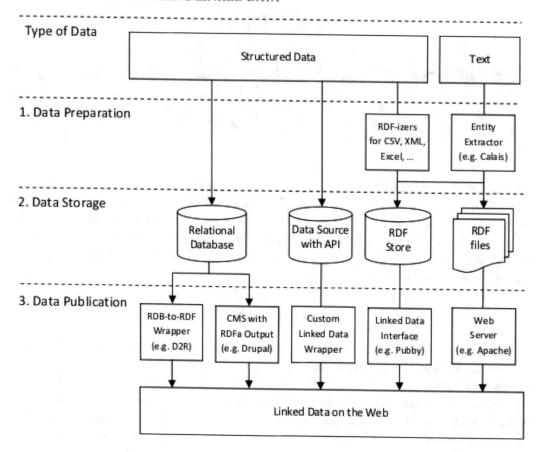

Figure 5.1: Linked Data Publishing Options and Workflows.

Where structured data exists in queryable form behind a custom API (such as the Flickr or Amazon Web APIs, or a local application or operating system API), the situation is a little more complex, as a custom wrapper will likely need to be developed according to the specifics of the API in question. However, examples such as the RDF Book Mashup [29] demonstrate that such wrappers can be implemented in relatively trivial amounts of code, much of which can likely be componentised for reuse across wrappers. The wrapper pattern is described in more detail in Section 5.2.6.

5.1.1.2   From Static Structured Data to Linked Data

Static input data may consist of CSV files, Excel spreadsheets, XML files or database dumps. In order to serve them as Linked Data on the Web, they must undergo a conversion process that outputs static RDF files or loads converted data directly into an RDF store. Lists of RDFizing tools that can be used for this conversion can be found at[1] and[2].

Where static files are already in RDF form and follow the Linked Data principles, they can simply be served up on the Web using a classic Web server (see Section 5.2.1) or loaded into an RDF store that has a suitable Linked Data interface (see Section 5.2.5).

5.1.1.3 From Text Documents to Linked Data

Where the input to a Linked Data publishing workflow comprises of textual documents in natural language, e.g., a series of news stories or business reports, it is possible to pass these documents through a Linked Data entity extractor such as Calais[3], Ontos[4] or DBpedia Spotlight[5] which annotate documents with the Linked Data URIs of entities referenced in the documents. Publishing these annotations together with the documents increases the discoverability of the documents and enables applications to use the referenced Linked Data sources as background knowledge to display complementary information on web pages or to enhance information retrieval tasks, for instance, offer faceted browsing instead of simple full-text search.

5.1.2 ADDITIONAL CONSIDERATIONS

Irrespective of how the data is currently stored, a number of additional factors have a bearing on the choice of publishing pattern:

5.1.2.1 Data Volume: How much data needs to be served?

The amount of data you want to serve will have a strong bearing on the publishing pattern you choose. If you only wish to publish a small amount of data, perhaps a few hundred RDF triples about one entity, then it is probably desirable to serve these as a static RDF file, as described in Recipe 5.2.1. This may necessitate more manual effort in data management if the same data must also be maintained in another location or format, but avoids the greater setup costs associated with more technically complex patterns.

To avoid wasting bandwidth and forcing browsers to load and parse very large RDF files, larger data sets that describe multiple entities should be split into separate files, usually one per entity described, that can be loaded as necessary. These can be served as static files, or loaded into an RDF store with a suitable Linked Data interface, as described in the recipes below.

5.1.2.2 Data Dynamism: How often does the data change?

In common with data volume, the rate of change in your data set will affect the publishing pattern you choose. Data that changes rarely, such as historical records, may be well suited to publication

---

[1]http://esw.w3.org/ConverterToRdf
[2]http://simile.mit.edu/wiki/RDFizers
[3]http://www.opencalais.com/
[4]http://www.ontos.com/o_eng/index.php?cs=1
[5]http://wiki.dbpedia.org/spotlight

as static files. These could be batch generated from an existing system and served by a conventional Web server configured appropriately (see Recipe 5.2.1).

If your data changes frequently it will be preferable to use a storage and management mechanism (such as an RDF store – see Section 5.2.5) that enables frequent changes. Where the data is already stored in a relational database, or other system that exposes an API, then an RDB-to-RDF wrapper (see Section 5.2.4) or custom wrapper (see Section 5.2.6), respectively, is likely to be preferable as it will minimise disruption to existing systems and workflows.

5.2   THE RECIPES

This section will discuss the different Linked Data publishing patterns in more detail, in the form of concrete publishing recipes.

5.2.1   SERVING LINKED DATA AS STATIC RDF/XML FILES

Producing static RDF files and uploading them to a Web server is probably the simplest way to publish Linked Data, and is a common approach when:

- a person creates and maintains relatively small RDF files manually, e.g., when publishing RDFS vocabularies or personal profiles in RDF

- a software tool or process generates or exports RDF data as static files

The majority of examples in this book is shown in the Turtle serialisation of RDF, for readability; however, if data is published using just one serialisation format, this should be RDF/XML, as it is widely supported by tools that consume Linked Data.

In the case of *Big Lynx*, serving a static RDF/XML file is the perfect recipe for publishing the company profile as Linked Data. The code sample below shows what this company profile looks like, converted to the Turtle serialisation of RDF.

```
1   @prefix rdf: <http://www.w3.org/1999/02/22-rdf-syntax-ns#> .
2   @prefix rdfs: <http://www.w3.org/2000/01/rdf-schema#> .
3   @prefix dcterms: <http://purl.org/dc/terms/> .
4   @prefix foaf: <http://xmlns.com/foaf/0.1/> .
5   @prefix sme: <http://biglynx.co.uk/vocab/sme#> .
6
7   <http://biglynx.co.uk/company.rdf#company>
8     rdf:type sme:SmallMediumEnterprise ;
9     foaf:name "Big Lynx Productions Ltd" ;
10    rdfs:label "Big Lynx Productions Ltd" ;
11    dcterms:description "Big Lynx Productions Ltd is an independent television
          production company based near Birmingham, UK, and recognised worldwide for
          its pioneering wildlife documentaries" ;
12    foaf:based_near <http://sws.geonames.org/3333125/> ;
13    sme:hasTeam <http://biglynx.co.uk/teams/management> ;
14    sme:hasTeam <http://biglynx.co.uk/teams/production> ;
15    sme:hasTeam <http://biglynx.co.uk/teams/web> ;
16
```

```
17   <http://biglynx.co.uk/teams/management> ;
18     rdf:type sme:Team ;
19     rdfs:label "The Big Lynx Management Team" ;
20     sme:leader <http://biglynx.co.uk/people/dave-smith> ;
21     sme:isTeamOf <http://biglynx.co.uk/company.rdf#company> .
22
23   <http://biglynx.co.uk/teams/production> ;
24     rdf:type sme:Team ;
25     rdfs:label "The Big Lynx Production Team" ;
26     sme:leader <http://biglynx.co.uk/people/matt-briggs> ;
27     sme:isTeamOf <http://biglynx.co.uk/company.rdf#company> .
28
29   <http://biglynx.co.uk/teams/web> ;
30     rdf:type sme:Team ;
31     rdfs:label "The Big Lynx Web Team" ;
32     sme:leader <http://biglynx.co.uk/people/nelly-jones> ;
33     sme:isTeamOf <http://biglynx.co.uk/company.rdf#company> .
```

5.2.1.1 Hosting and Naming Static RDF Files

This static profile document will be published at `http://biglynx.co.uk/company.rdf`. The fragment identifier `#company` is added to the URI of the document to give a URI for the company of `http://biglynx.co.uk/company.rdf#company`. While this static file references URIs in the *Big Lynx* namespace that use the *303 URI* pattern, the URI `http://biglynx.co.uk/company.rdf#company` is minted within the context of the document `http://biglynx.co.uk/company.rdf` and therefore must use the *hash URI* pattern.

Once it has been created, this static RDF/XML file can be uploaded to the *Big Lynx* Web server using FTP or any other method Nelly prefers. Those considering this approach should also refer to Section 5.2.5 for alternative methods for serving static files that are small enough to be kept in a Web server's main memory.

5.2.1.2 Server-Side Configuration: MIME Types

While serving static RDF/XML files is a very simple approach to publishing Linked Data, there are a number of *housekeeping* issues to be attended to. Older web servers may not be configured to return the correct *MIME type* when serving RDF/XML files, `application/rdf+xml`. They may instead return content using the MIME type `text/plain`, which can cause Linked Data applications to not recognise the content as RDF and therefore fail to process it.

The appropriate method for fixing this issue depends on the Web server being used. In the case of the widely deployed *Apache* Web server, the following line should be added to the `httpd.conf` configuration file, or to an `.htaccess` file in the directory on the server where the RDF files are placed:

```
1   AddType application/rdf+xml .rdf
```

This tells Apache to serve files with an `.rdf` extension using the correct MIME type for RDF/XML. This implies that files have to be named with the `.rdf` extension.

The following lines can also be added at the same time, to ensure the server is properly configured to serve RDF data in its N3 and Turtle serialisations[6]:

```
1   AddType text/n3;charset=utf-8 .n3
2   AddType text/turtle;charset=utf-8 .ttl
```

5.2.1.3   Making RDF Discoverable from HTML

With the company profile online and the *Big Lynx* Web server configured to serve the correct MIME type, Nelly is keen to ensure it is discoverable, not least because it provides links into the rest of the *Big Lynx* data set. A well-established convention for doing this involves using the `<link>` tag in the header of a related HTML document to point to the RDF file.

In the *Big Lynx* scenario there is an HTML page at `http://biglynx.co.uk/company.html` that describes the company. By adding the following line to the header of this HTML document, Nelly can help make the company profile in RDF visible to Web crawlers and Linked Data-aware Web browsers. This is known as the *Autodiscovery* pattern[7].

```
1   <link rel="alternate" type="application/rdf+xml" href="company.rdf">
```

It should be noted that this technique can be applied in all publishing scenarios and should be used throughout a Web site to aid discovery of data.

5.2.2   SERVING LINKED DATA AS RDF EMBEDDED IN HTML FILES

An alternative to publishing the *Big Lynx* company profile as RDF/XML is to embed it within the HTML page that describes the company, at `http://biglynx.co.uk/company.html`, using RDFa (introduced in Section 2.4.2). Nellymay choose to use this route to avoid having to manually update two static documents if the company information changes.

The example below shows how this could be achieved:

```
1    <?xml version="1.0" encoding="UTF-8"?>
2    <!DOCTYPE html PUBLIC "-//W3C//DTD XHTML+RDFa 1.0//EN"
3     "http://www.w3.org/MarkUp/DTD/xhtml-rdfa-1.dtd">
4    <html
5      xml:lang="en"
6      version="XHTML+RDFa 1.0"
7      xmlns="http://www.w3.org/1999/xhtml"
8      xmlns:rdf="http://www.w3.org/1999/02/22-rdf-syntax-ns#"
9      xmlns:rdfs="http://www.w3.org/2000/01/rdf-schema#"
10     xmlns:foaf="http://xmlns.com/foaf/0.1/"
11     xmlns:dcterms="http://purl.org/dc/terms/"
12     xmlns:sme="http://biglynx.co.uk/vocab/sme#"
13     >
14     <head>
15       <title>About Big Lynx Productions Ltd</title>
16       <meta property="dcterms:title" content="About Big Lynx Productions Ltd" />
```

---

[6]Guidance on correct media types for N3 and Turtle is taken from `http://www.w3.org/2008/01/rdf-media-types`
[7]`http://patterns.dataincubator.org/book/autodiscovery.html`

```
17    <meta property="dcterms:creator" content="Nelly Jones" />
18    <link rel="rdf:type" href="foaf:Document" />
19    <link rel="foaf:topic" href="#company" />
20    </head>
21    <body>
22    <h1 about="#company" typeof="sme:SmallMediumEnterprise" property="foaf:name"
          rel="foaf:based_near" resource="http://sws.geonames.org/3333125/">Big Lynx
          Productions Ltd</h1>
23    <div about="#company" property="dcterms:description">Big Lynx Productions Ltd
          is an independent television production company based near Birmingham, UK,
          and recognised worldwide for its pioneering wildlife documentaries</div>
24    <h2>Teams</h2>
25    <ul about="#company">
26      <li rel="sme:hasTeam">
27        <div about="http://biglynx.co.uk/teams/management" typeof="sme:Team">
28          <a href="http://biglynx.co.uk/teams/management"
               property="rdfs:label">The Big Lynx Management Team</a>
29          <span rel="sme:isTeamOf" resource="#company"></span>
30          <span rel="sme:leader"
               resource="http://biglynx.co.uk/people/dave-smith"></span>
31    </div>
32      </li>
33      <li rel="sme:hasTeam">
34        <div about="http://biglynx.co.uk/teams/production" typeof="sme:Team">
35          <a href="http://biglynx.co.uk/teams/production"
               property="rdfs:label">The Big Lynx Production Team</a>
36          <span rel="sme:isTeamOf" resource="#company"></span>
37          <span rel="sme:leader"
               resource="http://biglynx.co.uk/people/matt-briggs"></span>
38    </div>
39      </li>
40      <li rel="sme:hasTeam">
41        <div about="http://biglynx.co.uk/teams/web" typeof="sme:Team">
42          <a href="http://biglynx.co.uk/teams/web" property="rdfs:label">The Big
               Lynx Web Team</a>
43          <span rel="sme:isTeamOf" resource="#company"></span>
44          <span rel="sme:leader"
               resource="http://biglynx.co.uk/people/nelly-jones"></span>
45    </div>
46      </li>
47      </ul>
48    </body>
49    </html>
```

This RDFa produces the following Turtle output (reformatted slightly for readability) when passed through the *RDFa Distiller and Parser*[8]:

```
1    @prefix dcterms: <http://purl.org/dc/terms/> .
2    @prefix foaf: <http://xmlns.com/foaf/0.1/> .
3    @prefix rdf: <http://www.w3.org/1999/02/22-rdf-syntax-ns#> .
4    @prefix rdfs: <http://www.w3.org/2000/01/rdf-schema#> .
5    @prefix sme: <http://biglynx.co.uk/vocab/sme#> .
6
7    <http://biglynx.co.uk/company.html>
```

[8]http://www.w3.org/2007/08/pyRdfa/

```
8     a <foaf:Document> ;
9     dcterms:creator "Nelly Jones"@en ;
10    dcterms:title "About Big Lynx Productions Ltd"@en ;
11    foaf:topic <http://biglynx.co.uk/company.html#company> .
12
13    <http://biglynx.co.uk/company.html#company> a sme:SmallMediumEnterprise ;
14      sme:hasTeam
15        <http://biglynx.co.uk/teams/management>,
16        <http://biglynx.co.uk/teams/production>,
17        <http://biglynx.co.uk/teams/web> ;
18      dcterms:description "Big Lynx Productions Ltd is an independent television
               production company based near Birmingham, UK, and recognised worldwide for
               its pioneering wildlife documentaries"@en ;
19      foaf:based_near <http://sws.geonames.org/3333125/> ;
20      foaf:name "Big Lynx Productions Ltd"@en .
21
22    <http://biglynx.co.uk/teams/management>
23      a sme:Team ;
24      rdfs:label "The Big Lynx Management Team"@en ;
25      sme:isTeamOf <http://biglynx.co.uk/company.html#company> ;
26      sme:leader <http://biglynx.co.uk/people/dave-smith> .
27
28    <http://biglynx.co.uk/teams/production>
29      a sme:Team ;
30      rdfs:label "The Big Lynx Production Team"@en ;
31      sme:isTeamOf <http://biglynx.co.uk/company.html#company> ;
32      sme:leader <http://biglynx.co.uk/people/matt-briggs> .
33
34    <http://biglynx.co.uk/teams/web>
35      a sme:Team ;
36      rdfs:label "The Big Lynx Web Team"@en ;
37      sme:isTeamOf <http://biglynx.co.uk/company.html#company> ;
38      sme:leader <http://biglynx.co.uk/people/nelly-jones> .
```

Note how the URI identifying *Big Lynx* has changed to `http://biglynx.co.uk/company.html#company` because the URI of the document in which it is defined has changed.

RDFa can be particularly useful in situations where publishing to the Web makes extensive use of existing templates, as these can be extended to include RDFa output. This makes RDFa a common choice for adding Linked Data support to content management systems and Web publishing frameworks such as Drupal[9], which includes RDFa publishing support in version 7.

Care should be taken when adding RDFa support to HTML documents and templates, to ensure that the elements added produce the intended RDF triples. The complexity of this task increases with the complexity of the HTML markup in the document. Frequent use of the *RDFa Distiller and Parser*[10] and inspection of its output can help ensure the correct markup is added.

---

[9]http://drupal.org/
[10]http://www.w3.org/2007/08/pyRdfa/

5.2.3   SERVING RDF AND HTML WITH CUSTOM SERVER-SIDE SCRIPTS

In many Web publishing scenarios, the site owner or developer will have a series of custom server-side scripts for generating HTML pages and may wish to add Linked Data support to the site. This situation applies to the *Big Lynx* blogging software, which is powered by a series of custom PHP scripts that query a relational database and output the blog posts in HTML, at URIs such as `http://biglynx.co.uk/blog/making-pacific-sharks.html`.

Nelly considered enhancing these scripts to publish RDFa describing the blog posts, but was concerned that invalid markup entered in the body of blog posts by *Big Lynx* staff may make this data less consumable by RDFa-aware tools. Therefore, she decided to supplement the HTML-generating scripts with equivalents publishing Linked Data in RDF/XML. These scripts run the same database queries, and output the data in RDF/XML rather than HTML. This is achieved with the help of the ARC library for working with RDF in PHP[11], which avoids Nelly having to write an RDF/XML formatter herself. The resulting RDF documents are published at URIs such as `http://biglynx.co.uk/blog/making-pacific-sharks.rdf`.

A key challenge for Nelly at this stage is to ensure the RDF output can be classed as Linked Data, by including outgoing links to other resources within the *Big Lynx* data sets. This may involve mapping data returned from the relational database (e.g., names of productions or blog post authors) to known URI templates within the *Big Lynx* namespace.

To complete the process of Linked Data-enabling the *Big Lynx* blog, Nelly must make dereferenceable URIs for the blog posts themselves (as distinct from the HTML and RDF documents that describe the post). These URIs will take the form `http://biglynx.co.uk/blog/making-pacific-sharks`, as introduced in Section 4.3.3.

As these URIs follow the *303 URI* pattern (see Section 2.3.1), Nelly must create a script that responds to attempts to dereference these URIs by detecting the requested content type (specified in the `Accept:` header of the HTTP request) and performing a 303 redirect to the appropriate document. This is easily achieved using a scripting language such as PHP. Various code samples demonstrating this process are available at [12].

In fact, Nelly decides to use a `mod_rewrite`[13] rule on her server to catch all requests for blog entries and related documents, and pass these to one central script. This script then detects the nature of the request and either performs content negotiation and a 303 redirect, or calls the scripts that serve the appropriate HTML or RDF documents. This final step is entirely optional, however, and can be emitted in favour of more PHP scripts if `mod_rewrite` is not available on the Web server.

5.2.4   SERVING LINKED DATA FROM RELATIONAL DATABASES

There are many cases where data is stored in a relational database, perhaps powering an important legacy application, but would benefit from being exposed to the Web (or a corporate intranet)

---

[11]`http://arc.semsol.org/`
[12]`http://linkeddata.org/conneg-303-redirect-code-samples`
[13]`http://httpd.apache.org/docs/2.0/mod_mod_rewrite.html`

as Linked Data. In such cases it is generally advisable to retain the existing data management infrastructure and software, so as not to disrupt legacy applications, and instead simply publish a Linked Data view of the relational database.

One example of such a scenario is the *Big Lynx* job vacancies database, which drives the publication of past and present job vacancy information on the *Big Lynx* Web site in HTML. In this case, Nelly could write some additional server-side scripts to publish vacancies in parallel as Linked Data, as described above in 5.2.3. However, to explore alternatives to this approach, Nelly decided to use software that provides a Linked Data view over a relational database.

One widely used tool designed for this purpose is D2R Server[14]. D2R Server relies on a declarative mapping between the database schema and the target RDF terms, provided by the data publisher. Based on this mapping, D2R Server serves a SPARQL endpoint and Linked Data views of the database.

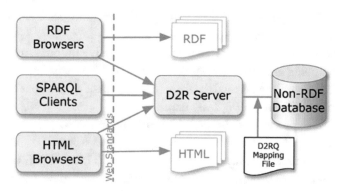

Figure 5.2: Architecture Diagram of D2R Server

Using D2R Server to publish a relational database as Linked Data typically involves the following steps:

1. Download and install the server software as described in the Quick Start[15] section of the D2R Server homepage.

2. Have D2R Server auto-generate a D2RQ mapping from the schema of your database.

3. Customize the mapping by replacing auto-generated terms with terms from well-known and publicly accessible RDF vocabularies (see Section 4.4.4).

4. Set RDF links pointing at external data sources as described in Section 4.5.

5. Set several RDF links from an existing interlinked data source (for instance, your FOAF profile) to resources within the new data set, to ensure crawlers can discover the data.

---

[14]http://sites.wiwiss.fu-berlin.de/suhl/bizer/d2r-server/index.html
[15]http://sites.wiwiss.fu-berlin.de/suhl/bizer/d2r-server/index.html#quickstart

6. Add the new data source to the CKAN registry in the group *LOD Cloud* as described in Section 3.1.

In addition to D2R Server, the following tools enable relational databases to be published as Linked Data:

- OpenLink Virtuoso[16] provides the *Virtuoso RDF Views*[17] Linked Data wrapper.

- Triplify[18] is a small plugin for Web applications, which allows you to map the results of SQL queries into RDF, JSON and Linked Data.

The W3C RDB2RDF Working Group[19] is currently working on a standard language to express relational database to RDF mappings. Once this language is finished, it might replace the solutions described above.

5.2.5   SERVING LINKED DATA FROM RDF TRIPLE STORES

Ideally, every RDF triple store software would provide a Linked Data interface. Using this interface, the administrator of the store could configure with part of the store's content should be made accessible as Linked Data on the Web.

However, for RDF stores where such interfaces are not yet available, Pubby[20] can act as a Linked Data interface in front of the triple store's SPARQL endpoint. Pubby rewrites URI-dereferencing requests into SPARQL DESCRIBE queries against the underlying RDF store and handles 303 redirects and content negotiation.

If a data set is sufficiently small, Pubby offers the `conf:loadRDF` option which loads RDF data from a static file and keeps it in the server's main memory. Thus, Pubby also provides a simple alternative to using Recipe 5.2.1 for serving static RDF files without needing to configure 303 redirects and content negotiation yourself.

The *ARC*[21] software library provides RDF storage, SPARQL querying and a Linked Data interface for *LAMP (Linux, Apache, MySQL, PHP)* environments which are commonly provided by cheap, conventional Web hosting companies.

5.2.6   SERVING LINKED DATA BY WRAPPING EXISTING APPLICATION OR WEB APIS

It is increasingly common for large Web sites such as Amazon[22], Flickr[23] and Twitter[24] to expose their data for reuse via *Web APIs*. A comprehensive list of such APIs can be found at Programmable

---

[16]http://virtuoso.openlinksw.com/wiki/main/Main/
[17]http://virtuoso.openlinksw.com/dataspace/dav/wiki/Main/VOSSQLRDF
[18]http://triplify.org/Overview
[19]http://www.w3.org/2001/sw/rdb2rdf/
[20]http://www4.wiwiss.fu-berlin.de/pubby/
[21]http://arc.semsol.org/
[22]http://www.amazon.com/
[23]http://flickr.com/
[24]http://twitter.com/

Web[25]. These various APIs provide rather heterogeneous query and retrieval interfaces and return results using a number of different formats such as XML, JSON or ATOM.

The content of the data stores behind these API can be made available as Linked Data by implementing wrappers around the APIs. In general, Linked Data wrappers do the following:

1. They assign HTTP URIs to the resources about which the API provides data.

2. When one of these URIs is dereferenced asking for `application/rdf+xml`, the wrapper rewrites the client's request into a request against the underlying API.

3. The results of the API request are transformed to RDF and sent back to the client.

This can be a simple and effective mechanism for exposing new sources of Linked Data. However, care should be taken to ensure adequate outgoing links are created from the wrapped data set, and that the individual or organisation hosting the wrapper has the rights to republish data from the API in this way.

5.3   ADDITIONAL APPROACHES TO PUBLISHING LINKED DATA

The primary means of publishing Linked Data on the Web is by making URIs dereferenceable, thereby enabling the follow-your-nose style of data discovery. This should be considered the minimal requirements for Linked Data publishing. In addition, Linked Data publishers may also provide RDF data set dumps for local replication of data, and SPARQL endpoints for querying the data directly. Providings such additional means of access anables Linked Data applications to choose the access method that best fits their needs. Mechanisms for advertising the availability of these are described in Section 4.3.1.

5.4   TESTING AND DEBUGGING LINKED DATA

During preparations for publishing Linked Data, data and publishing infrastructure should be checked to ensure it adheres to the Linked Data principles and best practices. A useful starting point for testing Linked Data is to check that RDF data conveys the intended information. A simple way to achieve this is to serialise the RDF as *N-Triples* (see Section 2.4.2) and *read* the data, checking that the triples themselves *make sense*. The W3C RDF Validator [26] can check RDF/XML for syntactic correctness, and provides tabular *N-Triples-like* output of validated triples that is useful for visual inspection, as described above. More in-depth analysis of data can be carried out using tools such as *Eyeball*[27].

[25]http://www.programmableweb.com/
[26]http://www.w3.org/RDF/Validator/
[27]http://jena.sourceforge.net/Eyeball/

An equally important step is to check the correct operation of hosting infrastructure. There are three public validation services available that check whether given URIs dereference correctly into RDF descriptions:

- The **Vapour Linked Data Validator** dereferences given URIs and provides a detailed report about the HTTP communication that tool place in the lookup process. The Vapour Linked Data Validator is available at `http://idi.fundacionctic.org/vapour`

- **RDF:Alerts** dereferences not only given URIs but also retrieves the definitions of vocabulary terms and checks whether data complies with `rdfs:range` and `rdfs:domain` as well as datatype restrictions that are given in these definitions. The RDF:Alerts validator is available at `http://swse.deri.org/RDFAlerts/`

- **Sindice Inspector** be used to visualize and validate RDF files, HTML pages embedding microformats, and XHTML pages embedding RDFa. The validator performs reasoning and checks for common errors as observed in RDF data found on the web. The Sindice Inspector is available at `http://inspector.sindice.com/`

Various other tools exist that enable more manual validation and debugging of Linked Data publishing infrastructure. The command line tool cURL[28] can be very useful in validating the correct operation of 303 redirects used with *303 URIs* (see Section 2.3.1), as described in the tutorial *Debugging Semantic Web sites with cURL*[29].

The Firefox browser extensions *LiveHTTPHeaders*[30] and *ModifyHeaders*[31] provide convenient GUIs for making HTTP requests with modified headers, and assessing the response from a server.

A more qualitative approach, that complements more technical debugging and validation, is to test with a data set can be fulled navigated using a Linked Data browser. For example, RDF links may be served that point from one resource to another, but not incoming links that connect the second resource back to the first. Consequently, using a Linked Data browser it may only be possible to navigate deeper into the data set but not return to the staring point. Testing for this possibility with a Linked Data browser will also highlight whether Linked Data crawlers can reach the entirety of the data set for indexing.

The following Linked Data browsers are useful starting points for testing:

Tabulator[32]. If Tabulator takes some time to display data, it may indicate that the RDF documents being served are too large, and may benefit from splitting into smaller fragments, or from omission of data that may be available elsewhere (e.g., triples describing resources which are not the primary subject of the document). Tabulator also performs some basic inferencing over data it consumes, without checking this for consistency. Therefore, unpredictable results

---

[28]`http://curl.haxx.se/`
[29]`http://dowhatimean.net/2007/02/debugging-semantic-web-sites-with-curl`
[30]`https://addons.mozilla.org/af/firefox/addon/3829/`
[31]`https://addons.mozilla.org/af/firefox/addon/967/`
[32]`http://www.w3.org/2005/ajar/tab`

when using this browser may indicate issues with `rdfs:subClassOf` and `rdfs:subPropertyOf` declarations in the RDFS and OWL schemas used in your data.

Marbles[33]. This browser uses a two second time-out when retrieving data from the Web. Therefore, if the browser does not display data correctly it may indicate that the host server is too slow in responding to requests.

References to further Linked Data browsers that can be used to test publishing infrastructure are given in Section 6.1.1. Alternatively, the *LOD Browser Switch*[34] can be used to dereference URIs from a data set within different Linked data browsers.

Beside validating that Linked Data is published correctly from a technical perspective, the data should be made as self-descriptive as possible, to maximise its accessibility and utility. The following section presents a checklist that can be used to validate that a data set complies with the various Linked Data best practices. An analysis of common errors and flaws of existing Linked Data sources is presented in [64] and provides a valuable source of negative examples.

5.5   LINKED DATA PUBLISHING CHECKLIST

In addition to providing your data via dereferenceable HTTP URIs, you data set should also comply with the best practices that ensure that data is as self-descriptive as possible, thereby enabling client applications to discover all relevant meta-information required to integrate data from different sources. The checklist below can be used by Linked Data publishers to verify that data meets the key requirements.

1. **Does your data set links to other data sets?** RDF links connect data from different sources into a single global RDF graph and enable Linked Data browsers and crawlers to navigate between data sources. Thus your data set should set RDF links pointing at other data sources [16].

2. **Do you provide provenance metadata?** In order to enable applications to be sure about the origin of data as well as to enable them to assess the quality of data, data source should publish provenance meta data together with the primary data (see Section 4.3).

3. **Do you provide licensing metadata?** Web data should be self-descriptive concerning any restrictions that apply to its usage. A common way to express such restrictions is to attach a data license to published data, as described in 4.3.3. Doing so is essential to enable applications to use Web data on a secure legal basis.

4. **Do you use terms from widely deployed vocabularies?** In order to make it easier for applications to understand Linked Data, data providers should use terms from widely deployed vocabularies to represent data wherever possible (see Section 4.4.4).

[33]`http://www5.wiwiss.fu-berlin.de/marbles/`
[34]`http://browse.semanticweb.org/`

5. **Are the URIs of proprietary vocabulary terms dereferenceable?** Data providers often define proprietary terms that are used in addition to terms from widely deployed vocabularies. In order to enable applications to automatically retrieve the definition of vocabulary terms from the Web, the URIs identifying proprietary vocabulary terms should be made dereferenceable [23].

6. **Do you map proprietary vocabulary terms to other vocabularies?** Proprietary vocabulary terms should be related to corresponding terms within other (widely used) vocabularies in order to enable applications to understand as much data as possible and to translate data into their target schemata, as described in Section 2.5.3.

7. **Do you provide data set-level metadata?** In addition to making instance data self-descriptive, it is also desirable that data publishers provide metadata describing characteristic of complete data sets, for instance, the topic of a data set and more detailed statistics, as described in 4.3.1.

8. **Do you refer to additional access methods?** The primary way to publish Linked Data on the Web is to make the URIs that identity data items dereferenceable into RDF descriptions. In addition, various LOD data providers have chosen to provide two alternative means of access to their data via SPARQL endpoints or provide RDF dumps of the complete data set. If you do this, you should refer to the access points in your voiD description as described in Section 5.3.

The *State of the LOD Cloud* document[35] provides statistics about the extent to which deployed Linked Data sources meet the guidelines given above.

---

[35]http://lod-cloud.net/state

CHAPTER 6

Consuming Linked Data

All data that is published on the Web, according to the Linked Data principles becomes part of a single, global data space. This chapter will discuss how applications use this Web of Data. In general, applications are built to exploit the following properties of the Linked Data architecture:

1. **Standardized Data Representation and Access.** Integrating Linked Data from different sources is easier compared to integrating data from proprietary Web 2.0 APIs, as Linked Data relies on a standardized data model and standardized data access mechanism, and as data is published in a self-descriptive fashion.

2. **Openness of the Web of Data.** The Linked Data architecture is open and enables the discovery of new data sources at runtime. This enables applications to automatically take advantage of new data sources as they become available on the Web of Data.

   This chapter is structured as follows: Section 6.1 gives an overview of deployed Linked Data applications. Section 6.2 outlines the main tasks and techniques for building a Linked Data application using the example of a simple mashup. Section 6.3 gives an overview of the different architectural patterns that are implemented by Linked Data applications and discusses the different tasks involved in Linked Data consumption. The concluding Section 6.4 examines how data integration efforts by data publishers, data consumers and third parties may complement each other to decrease heterogeneity on the Web of Data in an evolutionary fashion.

6.1 DEPLOYED LINKED DATA APPLICATIONS

This section gives an overview of deployed Linked Data applications. As the availability of Linked Data is a relatively recent phenomenon, the presented applications are mostly first generation applications and prototypes that will likely undergo significant evolution as lessons are learned from their development and deployment. Nevertheless, they already give an indication of what will be possible in the future as well as of the architectural patterns that are emerging in the field of Linked Data.

   Linked Data applications can be classified into two categories: generic applications and domain-specific applications. The following section gives an overview of applications from both categories without aiming to provide a complete listing. For more complete and up-to-date listings, please refer to the application-related project pages in the ESW LOD wiki[1] as well as to the W3C page *Semantic Web Case Studies and Use Cases*[2].

---

[1]http://esw.w3.org/SweoIG/TaskForces/CommunityProjects/LinkingOpenData#Project_Pages
[2]http://www.w3.org/2001/sw/sweo/public/UseCases/

6.1.1 GENERIC APPLICATIONS

Generic Linked Data applications can process data from any topical domain, for instance, library as well as life science data. There are two basic types of generic Linked Data applications: Linked Data browsers and Linked Data search engines.

6.1.1.1 Linked Data Browsers

Just as traditional Web browsers allow users to navigate between HTML pages by following hypertext links, Linked Data browsers allow users to navigate between data sources by following RDF links. For example, a user may view DBpedia's RDF description of the city of Bristol (UK), follow a *hometown* link to the description of the band Portishead (which originated in the city), and from there onward into RDF data from Freebase describing songs and albums by that band. The result is that a user may begin navigation in one data source and progressively traverse the Web by following RDF rather than HTML links.

The *Disco* hyperdata browser[3] follows this approach and can be seen as a direct application of the hypertext navigation paradigm to the Web of Data. Structured data, however, provides human interface opportunities and challenges beyond those of the hypertext Web. People need to be able to explore the links between single data items, but also to aggregate and powerfully analyze data in bulk [62].

The *Tabulator* browser[4] [105], for example, allows the user to traverse the Web of Data and expose pieces of it in a controlled fashion, in *outline mode*; to discover and highlight a pattern of interest; and then query for any other similar patterns in the data Web. The results of the query form a table that can then be analyzed with various conventional data presentation methods, such as faceted browsing, maps, timelines, and so on. Tabulator and *Marbles*[5] [7] are among the data browsers which track the provenance of data while merging data about the same thing from different sources. Figure 6.1 depicts the Marbles Linked Data browser displaying data about Tim Berners-Lee which has been merged from different sources. The colored marbles next to each fact in the figure refer to the data sources which contain that fact.

A recently released Linked Data browser is *LinkSailor*[6]. Besides displaying data in a tabular source view, LinkSailor applies display templates to automatically arrange data in a meaningful fashion that is appropriate to the nature of that data.

The LOD Browser Switch[7] enables a specific Linked Data URI to be rendered within different Linked data browsers.

---

[3]http://www4.wiwiss.fu-berlin.de/bizer/ng4j/disco/
[4]http://www.w3.org/2005/ajar/tab
[5]http://marbles.sourceforge.net/
[6]http://linksailor.com/
[7]http://browse.semanticweb.org/

Figure 6.1: The Marbles Linked Data browser displaying data about Tim Berners-Lee. The colored dots indicate the data sources from which data was merged.

6.1.1.2 Linked Data Search Engines

A number of search engines have been developed that crawl Linked Data from the Web by following RDF links, and provide query capabilities over aggregated data. These search engines integrate data from thousands of data sources and thus nicely demonstrate the advantages of the open, standards-based Linked Data architecture, compared to Web 2.0 mashups which rely on a fixed set of data sources exposing proprietary interfaces.

Search engines such as *Sig.ma*[8] [107], *Falcons*[9] [40], and *SWSE*[10] [55] provide keyword-based search services oriented towards human users and follow a similar interaction paradigm as existing market leaders such as Google and Yahoo. The user is presented with a search box into which they

---

[8]http://sig.ma/
[9]http://iws.seu.edu.cn/services/falcons/documentsearch/
[10]http://www.swse.org/

can enter keywords related to the item or topic in which they are interested, and the application returns a list of results that may be relevant to the query.

However, rather than simply providing links from search results through to the source documents in which the queried keywords are mentioned, Linked Data search engines provide richer interaction capabilities to the user which exploit the underlying structure of the data. For instance, Falcons enables the user to filter search results by class and therefore limit the results to show, for example, only persons or entities belonging to a specific subclass of person, such as athlete or politician. Sig.ma, Falcons and SWSE provide summary views of the entity the user selects from the results list, alongside additional structured data crawled from the Web and links to related entities.

The Sig.ma search engine applies vocabulary mappings to integrate Web data as well as specific display templates to properly render data for human consumption. Figure 6.2 shows the Sig.ma search engine displaying data about Richard Cyganiak that has been integrated from 20 data sources. Another interesting aspect of the Sig.ma search engine is that it approaches the data quality challenges that arise in the open environment of the Web by enabling its users to choose the data sources from which the user's aggregated view is constructed. By removing low quality data from their individual views, Sig.ma users collectively create ratings for data sources on the Web as a whole.

Figure 6.2: Sig.ma Linked Data search engine displaying data about Richard Cyganiak.

The **Fillmore** - Western Addition/NOPA - **San Francisco**, CA 🔍

☆☆☆☆☆ 752 reviews - Price range: $$

752 Reviews of The **Fillmore** "Last night we went to see Chris Isaak and it was our first time at the **Fillmore**. We could not have been any more delighted with ...

www.yelp.com/biz/the-**fillmore**-san-francisco - United States - Cached - Similar

The **Fillmore San Francisco** - The Fillmore Schedule | Eventful 🔍

View The **Fillmore's** upcoming event schedule and profile - **San Francisco**, CA. The **Fillmore**, also known as **Fillmore** Auditorium, is located in San ...

The Radiators - Farewell Tour! - 100th GAMH show!	Fri, Jan 7
3 NIGHTS! - An Evening With - Dark Star Orchestra	Fri, Jan 7
Bird by Bird - The Soft White Sixties - The Trophy Fire ...	Fri, Jan 7

eventful.com › San Francisco venues - Cached - Similar

Figure 6.3: Google search results containing structured data in the form of Rich Snippets.

A search engine that focuses on answering complex queries over Web data is *VisiNav*[11] [54]. Queries are formulated by the user in an exploratory fashion and can be far more expressive than queries that Google and Yahoo can currently answer. For instance, VisiNav answers the query *"give me the URLs of all blogs that are written by people that Tim Berners-Lee knows!"* with a list of 54 correct URLs. Google and Yahoo just return links to arbitrary web pages describing Tim Berners-Lee himself.

While Sig.ma, VisiNav, SWSE and Falcons provide search capabilities oriented towards humans, another breed of services have been developed to serve the needs of applications built on top of distributed Linked Data. These application-oriented indexes, such as *Sindice*[12] [108], *Swoogle*[13], and *Watson*[14] provide APIs through which Linked Data applications can discover RDF documents on the Web that reference a certain URI or contain certain keywords.

The rationale for such services is that each new Linked Data application should not need to implement its own infrastructure for crawling and indexing the complete Web of Data. Instead, applications can query these indexes to receive pointers to potentially relevant RDF documents which can then be retrieved and processed by the application itself. Despite this common theme, these services have slightly different emphases. Sindice is oriented to providing access to documents containing instance data, while the emphasis of Swoogle and Watson is on finding ontologies that provide coverage of certain concepts relevant to a query.

A service that goes beyond finding Web data but also helps developers to integrate Web data is *uberblic*[15], which acts as a layer between data publishers and data consumers. The service

---

[11]http://sw.deri.org/2009/01/visinav/
[12]http://sindice.com/
[13]http://swoogle.umbc.edu/
[14]http://kmi-web05.open.ac.uk/Overview.html
[15]http://uberblic.org/

consolidates and reconciles information into a central data repository, and provides access to this repository through developer APIs.

It is interesting to note that traditional search engines like Google and Yahoo[16] have also started to use structured data from the Web within their applications. Google crawls RDFa and microformat data describing people, products, businesses, organizations, reviews, recipes, and events. It uses the crawled data to provide richer and more structured search results to its users in the form of Rich Snippets[17]. Figure 6.3 shows part of the Google search results for *Fillmore San Francisco*. Below the title of the first result, it can be seen that Google knows about 752 reviews of the Fillmore concert hall. The second Rich Snippet contains a listing of upcoming concerts at this concert hall.

Not only does Google use structured data from the Web to enrich search results, it has also begun to use extracted data to directly answer simple factual questions[18]. As is shown in Figure 6.4, Google answers a query about the *birth date of Catherine Zeta-Jones* not with a list of links pointing at Web pages, but provides the actual answer to the user: 25 September 1969. This highlights how the major search engines have begun to evolve into answering engines which rely on structured data from the Web.

Catherine Zeta-Jones date of birth — 25 September 1969 - Feedback
According to wikipedia.org, imdb.com, talktalk.co.uk **and 4 others** - ⊞ Show sources

Figure 6.4: Google result answering a query about the birth date of Catherine Zeta-Jones.

6.1.2   DOMAIN-SPECIFIC APPLICATIONS

There are also various Linked Data applications that cover the needs of specific user communities.

Listings of Linked Data applications that contribute to increasing government transparency, by combining and visualizing government data, are found on the data.gov[19] and data.gov.uk[20] websites. One example of these applications is the *US Global Foreign Aid Mashup*[21] shown in Figure 6.5. The application pulls together spending data from the United States Agency for International Development (USAID), the Department of Agriculture, and the Department of State and combines the data with background information about countries from the CIA World Factbook as well as with news articles from the New York Times. By combining the data, the mashup enables its users to recognize the current focus areas of US foreign aid as well as to analyze shifts in the geographic focus over time.

---

[16]http://linkeddata.future-internet.eu/images/5/54/Mika_FI_Search_and_LOD.pdf
[17]http://googlewebmastercentral.blogspot.com/2009/10/help-us-make-web-better-update-on-rich.html
[18]http://googleblog.blogspot.com/2010/05/understanding-web-to-find-short-answers.html
[19]http://www.data.gov/communities/node/116/apps
[20]http://data.gov.uk/apps
[21]http://data-gov.tw.rpi.edu/demo/USForeignAid/demo-1554.html

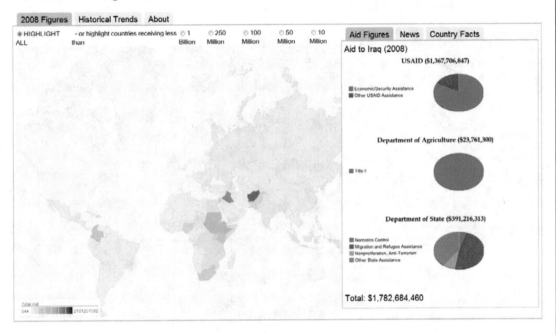

Figure 6.5: US Global Foreign Aid Mashup combining and visualizing data from different branches of the US government.

Linked Data applications that aim to bring Linked Data into the user's daily work context include *dayta.me* [3] and *paggr* [90]. dayta.me[22] is a personal information recommender that augments a person's online calendar with useful information pertaining to their upcoming activities. paggr[23] provides an environment for the personalized aggregation of Web data through dashboards and widgets.

An application of Linked Data that helps educators to create and manage lists of learning resources (e.g., books, journal articles, Web pages) is *Talis Aspire*[24] [41]. The application is written in PHP, backed by the Talis Platform[25] for storing, managing and accessing Linked Data, and used by tens of thousands of students at numerous universities on a daily basis. Educators and learners interact with the application through a conventional Web interface, while the data they create is stored natively in RDF. An HTTP URI is assigned to each resource, resource list, author and publisher. Use of the Linked Data principles and related technologies in Aspire enables individual

---

[22]http://dayta.me/
[23]http://paggr.com/
[24]http://www.w3.org/2001/sw/sweo/public/UseCases/Talis/
[25]http://www.talis.com/platform/

lists and resources to be connected to related data elsewhere on the Web, enriching the range of material available to support the educational process.

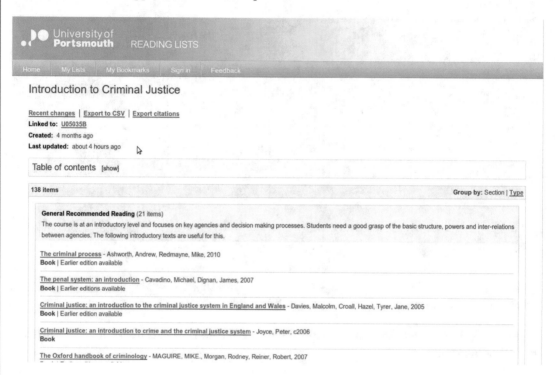

Figure 6.6: The HTML view of a Talis Aspire List generated from the underlying RDF representation of the data.

DBpedia Mobile[26] [7] is a Linked Data application that helps tourists to explore a city. The application runs on an iPhone or other smartphone and provides a location-centric mashup of nearby locations from DBpedia, based on the current GPS position of the mobile device. Using these locations as starting points, the user can then navigate along RDF links into other data sources. Besides accessing Web data, DBpedia Mobile also enables users to publish their current location, pictures and reviews to the Web as Linked Data, so that they can be used by other applications. Instead of simply being tagged with geographical coordinates, published content is interlinked with a nearby DBpedia resource and thus contributes to the overall richness of the Web of Data.

A Life Science application that relies on knowledge from more than 200 publicly available ontologies in order to support its users in exploring biomedical resources is the *NCBO Resource Index*[27] [69]. A second example of a Linked Data application from this domain is *Diseasome Map*[28].

---

[26] http://wiki.dbpedia.org/DBpediaMobile
[27] http://bioportal.bioontology.org/resources
[28] http://diseasome.eu/map.html

The application combines data from various Life Science data sources in order to generate a "network of disorders and disease genes linked by known disorder–gene associations, indicating the common genetic origin of many diseases."[29]

A Linked Data mashup that demonstrates how specific FOAF profiles are discovered and integrated is Researcher Map[30]. The application discovers the personal profiles of German database professors by following RDF links and renders the retrieved data on an interactive map [59].

A social bookmarking tool that allows tagging of bookmarks with Linked Data URIs to prevent ambiguities is Faviki[31]. Identifiers are automatically suggested using the Zemanta API[32], and Linked Data sources such as DBpedia and Freebase are used as background knowledge to organize tags by topics and to provide tag descriptions in different languages.

Applications that demonstrate how Linked Data is used within wiki-environments include Shortipedia[33] [112] and the Semantic MediaWiki - Linked Data Extension [34] [8].

6.2   DEVELOPING A LINKED DATA MASHUP

As an initial starting point for developing Linked Data applications, this section gives an overview of the main steps that are involved in developing a simple Linked Data Mashup. The use case for the mashup will be to augment the *Big Lynx* Web site with background information from the Web of Data about the places where *Big Lynx* employees live. The information should be displayed next to the information about the employee on his profile page. The mashup will be developed using the Java programming language and will conduct the following three steps:

1. Discover data sources that provide data about a city by following RDF links from an initial seed URI into other data sources.

2. Download data from the discovered data sources and store the data together with provenance meta-information in a local RDF store.

3. Retrieve information to be displayed on the *Big Lynx* Web site from the local store, using the SPARQL query language.

This simple example leaves out many important aspects that are involved in Linked Data consumption. Therefore, after explaining how the simple example is realized, an overview will be provided of the more complex tasks that need to be addressed by Linked Data applications (Section 6.3).

---

[29]http://diseasome.eu/poster.html
[30]http://researchersmap.informatik.hu-berlin.de/
[31]http://www.faviki.com/
[32]http://www.zemanta.com/
[33]http://shortipedia.org/
[34]http://smwforum.ontoprise.com/smwforum/index.php/SMW+LDE

6.2.1   SOFTWARE REQUIREMENTS

In this example mashup, two open-source tools will be used to access the Web of Data and to cache retrieved data locally for further processing:

1. **LDspider** [65], a Linked Data crawler that can process a variety of Web data formats including RDF/XML, Turtle, Notation 3, RDFa and many microformats. LDspider supports different crawling strategies and allows crawled data to be stored either in files or in an RDF store (via SPARQL/Update[35]).

2. **Jena TDB**, an RDF store which allows data to be added using SPARQL/Update and provides for querying the data afterwards using the SPARQL query language.

   LDspider can be downloaded from `http://code.google.com/p/ldspider/`. Use of LDspider requires a Java runtime environment on the host machine and inclusion of the LDspider `.jar` file in the machine's classpath.

   Jena TDB can be downloaded from `http://openjena.org/TDB/`. The site also contains instructions on how to install TDB. For the example mashup, the TBD standard configuration will be used and the store will be located at `localhost:2020`.

6.2.2   ACCESSING LINKED DATA URIS

The basic mechanism to access Linked Data on the Web is to dereference HTTP URIs into RDF descriptions and to follow RDF links from within the retrieved data into other data sources in order to discover additional related data.

   Dave Smith lives in Birmingham and the RDF data that *Big Lynx* maintains about him says that he is `foaf:based_near http://dbpedia.org/resource/Birmingham`. In order to get background information about Birmingham , this URI should be dereferenced and `owl:sameAs` links from the retrieved RDF followed, to a depth of one step into the Web of Data. Retrieved data should be stored within the RDF store. LDspider can be instructed to do this by issuing the following command on the command line.

```
1  java -jar ldspider.jar
2     -u "http://dbpedia.org/resource/Birmingham"
3     -b 5 10000
4     -follow "http://www.w3.org/2002/07/owl/sameAs"
5     -oe "http://localhost:2020/update/service"
```

The `-u` parameter provides LDspider with the DBpedia Birmingham URI as seed URI. The `-follow` parameter instructs LDspider to follow only `owl:sameAs` links and to ignore other link types. `-b` restricts the depth to which links are followed. The `-oe` parameter tells LDspider to put retrieved data via SPARQL/Update into the RDF store available at the given URI.

   LDspider starts with dereferencing the DBpedia URI. Within the retrieved data, LDspider discovers several `owl:sameAs` links pointing at further data about Birmingham provided by Geonames,

---

[35]`http://www.w3.org/TR/sparql11-update/`

Freebase, and the New York Times. In the second step, LDspider dereferences these URIs and puts the retrieved data into the RDF local store using SPARQL/Update.

6.2.3   REPRESENTING DATA LOCALLY USING NAMED GRAPHS

Linked Data applications regularly need to represent retrieved data together with provenance meta-information locally. A data model that is widely used for this task is *Named Graphs* [39]. The basic idea of Named Graphs is to take a set of RDF triples (i.e., a graph) and name this graph with a URI reference. Multiple Named Graphs can be represented together in the form of a graph set. As the graphs are identified with URI references, it is possible to talk about them in RDF, for instance, by adding triples to the graph itself which describe the creator or the retrieval date of the graph. Within these triples, the URI identifying the graph is used in the subject position.

The listing below uses the TriG Syntax[36], a simple extension to the Turtle syntax (see Section 2.4.2), that provides for representing sets of Named Graphs. Within TriG, each Named Graph is preceded by its name. The RDF triples that make up the graph are enclosed with curved brackets.

The listing begins on Line 1 by specifying the graph name: `http://localhost/myGraphNumberOne`. Lines 3 and 4 contain two RDF triples describing a restaurant. Lines 6 and 7 contain meta-information about the graph and state that the graph was created by Chris Bizer on December 17th, 2010.

```
1   <http://localhost/myGraphNumberOne>
2   {
3      biz:JoesPlace  rdfs:label  "Joe's Noodle Place"@en .
4      biz:JoesPlace  rev:rating  rev:excellent .
5
6      <http://localhost/myGraphNumberOne>  dc:creator
           <http://www4.wiwiss.fu-berlin.de/is-group/resource/persons/Person4> .
7      <http://localhost/myGraphNumberOne  dc:date> "2010-12-17"^^xsd:date .
8   }
```

LDspider uses the Named Graphs data model to store retrieved data. After LDspider has finished its crawling job, the RDF store contains four Named Graphs. Each graph is named with the URI from which LDspider retrieved the content of the graph. The listing below shows a subset of the retrieved data from DBpedia, Geonames, and the New York Times. The graph from Freebase is omitted due to space restrictions. The RDF store now contains a link to an image depicting Birmingham (Line 5) provided by DBpedia, geo-coordinates for Birmingham provided by Geonames (Lines 16 and 17), as well as a link that we can use to retrieve articles about Birmingham from the New York Times archive (Lines 26 and 27).

```
1   <http://dbpedia.org/data/Birmingham.xml>
2   {
3      dbpedia:Birmingham  rdfs:label  "Birmingham"@en .
4      dbpedia:Birmingham  rdf:type  dbpedia-ont:City .
5      dbpedia:Birmingham  dbpedia-ont:thumbnail
           <http://.../200px-Birmingham_-UK_-Skyline.jpg> .
```

---

[36]http://www4.wiwiss.fu-berlin.de/bizer/TriG/

```
6      dbpedia:Birmingham  dbpedia−ont:elevation  "140"^^xsd:double  .
7      dbpedia:Birmingham  owl:sameAs  <http://data.nytimes.com/N35531941558043900331>  .
8      dbpedia:Birmingham  owl:sameAs  <http://sws.geonames.org/3333125/>  .
9      dbpedia:Birmingham  owl:sameAs
            <http://rdf.freebase.com/ns/guid.9202...f8000000088c75>  .
10   }
11
12   <http://sws.geonames.org/3333125/about.rdf>
13   {
14     <http://sws.geonames.org/3333125/>  gnames:name  "City  and  Borough  of
            Birmingham" .
15     <http://sws.geonames.org/3333125/>  rdf:type  gnames:Feature  .
16     <http://sws.geonames.org/3333125/>  geo:long  "−1.89823" .
17     <http://sws.geonames.org/3333125/>  geo:lat  "52.48048" .
18     <http://sws.geonames.org/3333125/>  owl:sameAs
19       <http://www.ordnancesurvey.co.uk/...#birmingham_00cn>  .
20   }
21
22   <http://data.nytimes.com/N35531941558043900331>
23   {
24     nyt:N35531941558043900331  skos:prefLabel  "Birmingham  (England)"@en  .
25     nyt:N35531941558043900331  nyt:associated_article_count  "3"^^xsd:integer  .
26     nyt:N35531941558043900331  nyt:search_api_query
27       <http://api.nytimes.com/svc/search/...>  .
28   }
```

6.2.4 QUERYING LOCAL DATA WITH SPARQL

The SPARQL query language [95] is widely used for querying RDF data and is implemented by all major RDF stores. Besides querying single RDF graphs, SPARQL also provides for querying sets of Named Graphs. The language construct for querying multiple graphs are explained in detail in Section 8 of the SPARQL Recommendation[37].

The mashup wants to display data about Birmingham next to Dave's profile on the *Big Lynx* Web site. In order to retrieve all information that LDspider has found in all data sources about Birmingham , the mashup would execute the following SPARQL query against the RDF store. The mashup and the store will use the SPARQL Protocol[38] to exchange queries and query results.

```
1   SELECT DISTINCT ?p ?o ?g WHERE
2   {
3     { GRAPH ?g
4       { <http://dbpedia.org/resource/Birmingham> ?p ?o . }
5     }
6   UNION
7     { GRAPH ?g1
8     { <http://dbpedia.org/resource/Birmingham>
9       <http://www.w3.org/2002/07/owl#sameAs> ?y }
10    GRAPH ?g
11    { ?y ?p ?o }
12  } }
```

[37]http://www.w3.org/TR/rdf-sparql-query/#rdfDataset
[38]http://www.w3.org/TR/rdf-sparql-protocol/

The tokens starting with a question mark in the query are variables that are bound to values form the different graphs during query execution. The first line of the query specifies that we want to retrieve the predicates of all triples (?p) as well as the objects (?o) of all triples that describe Birmingham . In addition, we want to retrieve the names of the graphs (?g) from which each triple originates. We want to use the graph name to group triples on the web page and to display the URI from where a triple was retrieved next to each triple. The graph pattern in Lines 3-5 matches all data about Birmingham from DBpedia. The graph patterns in Lines 7-10 match all triples in other graphs that are connected by `owl:sameAs` links with the DBpedia URI for Birmingham .

Jena TDB sends the query results back to the application as a SPARQL result set XML document [39] and the application renders them to fit the layout of the web page.

The minimal Linked Data application described above leaves out many important aspects that are involved in Linked Data consumption. These are discussed below.

6.3   ARCHITECTURE OF LINKED DATA APPLICATIONS

This section gives an overview of the different architectural patterns that are implemented by Linked Data applications. This is followed by details of the different tasks that are involved in Linked Data consumption, with references to tools that can be used to handle these tasks within an application.

The architectures of Linked Data applications are very diverse and largely depend on the concrete use case. In general, however, one can distinguish the following three architectural patterns:

1. **The Crawling Pattern**. Applications that implement this pattern crawl the Web of Data in advance by traversing RDF links. Afterwards, they integrate and cleanse the discovered data and provide the higher layers of the application with an integrated view on the original data. The crawling pattern mimics the architecture of classical Web search engines like Google and Yahoo. The crawling pattern is suitable for implementing applications on top of an open, growing set of sources, as new sources are discovered by the crawler at run-time. Separating the tasks of building up the cache and using this cache later in the application context enables applications to execute complex queries with reasonable performance over large amounts of data. The disadvantage of the crawling pattern is that data is replicated and that applications may work with stale data, as the crawler might only manage to re-crawl data sources at certain intervals. The crawling pattern is implemented by the Linked Data search engines discussed in Section 6.1.1.2.

2. **The On-The-Fly Dereferencing Pattern** is implemented by Linked Data browsers discussed in Section 6.1.1.1. Within this pattern, URIs are dereferenced and links are followed the moment the application requires the data. The advantage of this pattern is that applications never process stale data. The disadvantage is that more complex operations are very slow as they might involve dereferencing thousands of URIs in the background. [57] propose an architecture for answering complex queries over the Web of Data that relies on the on-the-fly

---

[39] http://www.w3.org/TR/rdf-sparql-XMLres/

dereferencing pattern. As the results of this work show, data currency and a very high degree of completeness are achieved at the price of very slow query execution.

3. **The Query Federation Pattern** relies on sending complex queries (or parts of complex queries) directly to a fixed set of data sources. The pattern can be used if data sources provide SPARQL endpoints in addition to serving their data on the Web via dereferenceable URIs. The pattern enables applications to work with current data without needing to replicate complete data sources locally. The drawbacks of the federation approach have been extensively studied in the database community [72]. A major problem is that finding performant query execution plans for join queries over larger numbers of data sources is complex. As a result, the query performance is likely to slow down significantly when the number of data sources grows. Thus, the query federation pattern should only be used in situations where the number of data sources is known to be small. In order to provide for data source discovery within this query federation pattern, applications could follow links between data sources, examine voiD descriptions (see Section 4.3.1.2) provided by these data sources and then include data sources which provide SPARQL endpoints into their list of targets for federated queries.

Hartig and Langegger present a deeper comparison of the advantages and disadvantages of the architectural patterns in [58]. The appropriate pattern (or mixture of these patterns) will always depend on the specific needs of a Linked Data application. The factors that determine the decision for a specific pattern are:

1. the number of data sources that an application intends to use,

2. the degree of data freshness that is required by the application,

3. the required response time for queries and user interactions,

4. the extent to which the applications aims to discover new data sources at runtime.

However, due to the likelihood of scalability problems with on-the-fly link traversal and federated querying, it may transpire that widespread crawling and caching will become the norm in making data from a large number of data sources available to applications with acceptable query response times, while being able to take advantage of the openness of the Web of Data by discovering new data sources through link traversal.

Figure 6.7 gives an overview of the architecture of a Linked Data application that implements the crawling pattern. All data that is published on the Web, according to the Linked Data principles, becomes part of a giant global graph. This logical graph is depicted in the *Web of Linked Data Layer* in the lower part of Figure 6.7. Applications that implement the crawling pattern, but also applications that rely on the other patterns, typically implement the modules shown in the *Data Access, Integration and Storage Layer*. An overview of the tasks handled by these modules is given below:

1. Accessing the Web of Data. The basic means to access Linked Data on the Web is to dereference HTTP URIs into RDF descriptions and to discover additional data sources by traversing RDF

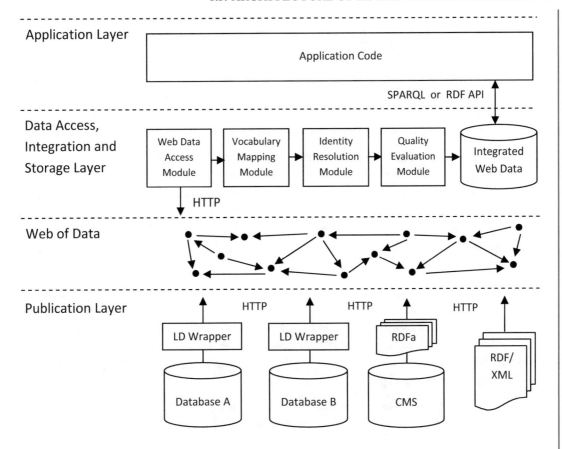

Figure 6.7: Architecture of a Linked Data application that implements the crawling pattern.

links. In addition, relevant data can also be discovered via Linked Data search engines and might be accessed via SPARQL endpoints or in the form of RDF data dumps.

2. **Vocabulary Mapping.** Different Linked Data sources may use different RDF vocabularies to represent the same type of information. In order to understand as much Web data as possible, Linked Data applications translate terms from different vocabularies into a single target schema. This translation may rely on vocabulary links that are published on the Web by vocabulary maintainers, data providers or third parties. Linked Data applications which discover data that is represented using terms that are unknown to the application may therefore search the Web for mappings and apply the discovered mappings to translate data to their local schemata.

3. **Identity Resolution.** Different Linked Data sources use different URIs to identify the same entity, for instance, a person or a place. Data sources may provide owl:sameAs links pointing

at data about the same real-world entity provided by other data sources. In cases where data sources do not provide such links, Linked Data applications may apply identity resolution heuristics in order to discover additional links.

4. Provenance Tracking. Linked Data applications rely on data from open sets of data sources. In order to process data more efficiently, they often cache data locally. For cached data, it is important to keep track of data provenance in order to be able to assess the quality of the data and to go back to the original source if required.

5. Data Quality Assessment. Due to the open nature of the Web, any Web data needs to be treated with suspicion, and Linked Data applications should thus consider Web data as claims by different sources rather than as facts. Data quality issues might not be too relevant if an application integrates data from a relatively small set of known sources. However, in cases where applications integrate data from the open Web, applications should employ data quality assessment methods in order to determine which claims to accept and which to reject as untrustworthy.

6. Using the Data in the Application Context. After completing tasks 1 to 5, the application has integrated and cleansed Web data to an extent that is required for more sophisticated processing. Such processing may in the most simple case involve displaying data to the user in various forms (tables, diagrams, other interactive visualizations). More complex applications may aggregate and/or mine the data, and they may employ logical reasoning in order to make implicit relationships explicit.

The following sections describe these tasks outlined above in more detail, and they refer to relevant papers and open-source tools that can be used to perform the tasks.

6.3.1 ACCESSING THE WEB OF DATA

The specific method to access the Web of Data depends on the architectural pattern implemented by an application. The basic means to access and navigate the graph is to dereference HTTP URIs into RDF descriptions and to traverse RDF links discovered within the retrieved data. In addition, parts of the graph may also be accessed via SPARQL endpoints or downloaded in the form of RDF data set dumps. Linked Data search engines cache the Web of Data and provide APIs to access the cached data. Therefore, instead of directly accessing the original data sources, applications can also access the Web of Data via the APIs provided by these search engines.

1. Linked Data Crawlers follow RDF links from a given set of seed URIs and store the retrieved data either in an RDF store or as local files. Several Linked Data search engines develop their own crawlers but do not open source them. A publicly available Linked Data crawler is *LDspider* [65] which was introduced in Section 6.2.2.

2. Linked Data Client Libraries support applications to dereference URIs into RDF data and provide for answering more complex queries by dereferencing multiple URIs on-the-fly [57].

Client libraries are available for different programming languages: JavaScript programmers can use the *Tabulator AJAR* library[40], PHP programmers *Moriarty*[41], Java programmers the *Semantic Web Client Library*[42]. The *SQUIN*[43] query service can be used within LAMP architectures.

3. **SPARQL Client Libraries** support applications in querying remote SPARQL endpoints over the SPARQL protocol. A SPARQL client library for PHP programmers is included in the *ARC* package[44], a Java library is provided as part of the *Jena* Semantic Web framework for Java[45].

4. **Federated SPARQL engines** include *DARQ* [96] and *SemaPlorer* [99]. The W3C SPARQL 1.1 working group is currently extending SPARQL with basic support for federated queries[46]. Once this extension is finished, it is likely that more RDF tools will start providing support for federated queries.

5. **RDFa Tools** extract RDF triples from HTML pages and provide for further processing them. A list of available tools is maintained in the RDFa wiki[47].

6. **Search Engine APIs** *Sindice*[48] as well as *Falcons*[49] provide APIs to access the data that they have crawled from the Web. An API that is specialized in owl:sameAs links is provided by *sameAs.org*[50]. An API that does not only support search but also consolidates Web data is provided by *uberblic*[51].

A an up-to-date list of Linked Data access tools is maintained by the LOD community in the ESW wiki[52]. A data snapshot available for experimentation without first needing to crawl the Web of Data is the Billion Triples Challenge (BTC) Dataset[53], provided for use by the participants of the Semantic Web Challenge and is updated at yearly intervals. The BTC 2010 data set was crawled in April 2010 and consists of 3.2 billion RDF triples (gzipped 27 GByte).

6.3.2   VOCABULARY MAPPING

Different Linked Data sources often use different vocabularies to represent data about the same type of entity [22]. In order to present a clean and integrated view on the data to their users, Linked Data

---

[40]http://dig.csail.mit.edu/2005/ajar/ajaw/Developer.html
[41]http://code.google.com/p/moriarty/
[42]http://www4.wiwiss.fu-berlin.de/bizer/ng4j/semwebclient/
[43]http://squin.org
[44]http://arc.semsol.org/
[45]http://jena.sourceforge.net/
[46]http://www.w3.org/TR/sparql11-federated-query/
[47]http://rdfa.info/wiki/Tools
[48]http://sindice.com/developers/api
[49]http://ws.nju.edu.cn/falcons/api/index.jsp
[50]http://sameas.org/about.php
[51]http://uberblic.org/
[52]http://esw.w3.org/TaskForces/CommunityProjects/LinkingOpenData/SemWebClients
[53]http://www.cs.vu.nl/~pmika/swc/submissions.html

applications may translate data from different vocabularies into the application's target schema. This translation can rely on vocabulary links, such as `owl:equivalentClass` and `owl:equivalentProperty` mappings as well as `rdfs:subClassOf` and `rdfs:subPropertyOf` statements that are published on the Web by vocabulary maintainers or data providers. The translation may also be based on additional mappings that are manually created or data-mined on the client-side.

RDF Schema and OWL provide terms for representing basic correspondences between vocabulary terms. They do not provide for more complex transformations, such as *structural transformations* like merging two resources into a single resource, or *property value transformations*, like splitting string values or normalizing units of measurement. Such features are provided by more expressive mapping languages such as SPARQL++ [94], the Alignment API [48], the Rules Interchange Format (RIF) and the mapping languages proposed by Haslhofer in [60]. A mapping framework that supports the publication and discovery of expressive mapping on the Web is the R2R Famework [34][54]. [42] presents a framework that uses mappings to rewrite SPARQL queries in a federated setting.

A data integration tool that can be used to manually create mappings is *Google Refine*[55] (RDF extension available from [56]). The *OpenII Framework* implements advanced schema clustering methods and can be used to support the mapping creation process [100].

A good overview of methods to automatically or semi-automatically generate mappings is presented by Euzenat & Shvaiko in [49]. Methods to machine learn mappings from instance data are presented by Nikolov & Motta in [88] and by Bilke & Naumann in [24]. Such instance-based methods are likely to produce good results in the Linked Data context as large amounts of instance data from the Web can be used for training.

6.3.3 IDENTITY RESOLUTION

Different Linked Data sources use different URIs to identify the same entity. Data sources may provide `owl:sameAs` links pointing at data about the same entity within other data sources. In cases where data sources do not provide such links, Linked Data applications may apply identity resolution heuristics in order to add additional links. In order to simplify local data processing, Linked Data applications often resolve URI aliases and locally represent all data that is available about an entity using a single (local) URI. A common identity resolution approach is to normalize URIs based on `owl:sameAs` statements contained in the data. In addition, applications can normalize identifiers based on `owl:InverseFunctionalProperty` values or employ more complex identity resolution heuristics [46], similar to the heuristics discussed in Section 4.5 on *Auto-generating RDF Links*.

An open-source tool that can be used as an identity resolution module within Linked Data applications is *Silk Server* [66]. Silk Server is designed to be used with an incoming stream of RDF instances, produced, for example, by a Linked Data crawler. Silk Server matches incoming instances against a local set of known instances and discovers duplicate instances. Based on this assessment,

[54] http://www4.wiwiss.fu-berlin.de/bizer/r2r/
[55] http://code.google.com/p/google-refine/
[56] http://lab.linkeddata.deri.ie/2010/grefine-rdf-extension/

an application can store data about newly discovered instances in its repository or fuse data that is already known about an entity with additional data from the Web.

6.3.4   PROVENANCE TRACKING

It is important for Linked Data applications to keep track of data provenance in order to be able to go back to the original source if required or to compare the quality of data from different sources. Many Linked Data applications employ the Named Graphs data model [39] together with the SPARQL query language to represent and query Web data together with provenance information as a local integrated model. This usage of Named Graphs has been described in Section 6.2. Different RDF vocabularies that can be used to represent provenance information were discussed in 4.3.2.

6.3.5   DATA QUALITY ASSESSMENT

Linked Data might be outdated, imprecise, or simply wrong. Therefore, Linked Data applications should consider all RDF statements that they discover on the Web as claims by a specific source rather than as facts. Applications should contain a module to filter RDF spam and prefer data from sources that are known for good quality to data from others. Data quality issues might not be too relevant, if an application integrates data only from a relatively small set of known sources. In cases where applications integrate data from the open Web, data quality assessment becomes crucial.

The specific methods that are used to assess the quality of Web data depend on the specific application context. In general, data quality assessment heuristics can be classified into three categories according to the type of information that is used as quality indicator [28]:

1. **Content-based Heuristics** use information to be assessed itself as quality indicator. The metrics analyze the information content or compare information with related information. Examples include outlier detection methods, for instance, treat a sales offer with suspicion if the price is more than 30% below the average price for the item, as well as classic spam detection methods that rely on patterns of suspicious words.

2. **Context-based Heuristics** Context-based metrics employ meta-information about the information content and the circumstances in which information was created, e.g., who said what and when, as quality indicator. A simple context-based heuristic is to prefer more recent data to older data. More complex heuristics could, for example, be to prefer product descriptions published by the manufacturer over descriptions published by a vendor or to disbelieve everything a vendor says about its competitor.

3. **Rating-based Heuristics** rely on explicit ratings about information itself, information sources, or information providers. Ratings may originate from the information consumer, other information consumers (as, for instance, the ratings gathered by the Sig.ma search engine, see Section 6.1.1.2), or domain experts.

Once the application has assessed the quality of a piece of information, it has different options for handling data conflicts and low quality data. Depending on the context, it is preferable to:

1. **Rank Data.** The simplest approach is to display all data, but rank data items according to their quality score. This approach is currently used by the Linked Data search engines discussed in Section 6.1.1.2. Inspired by Google, the search engines rely on variations of the PageRank algorithm [91] to determine coarse-grained measures of the popularity or significance of a particular data source, as a proxy for relevance or quality of the data.

2. **Filter Data.** In order to avoid overwhelming users with low-quality data, applications may decide to display only data which successfully passed the quality evaluation. A prototypical software framework that can be used to filter Web data using a wide range of different data quality assessment policies is the WIQA framework[57].

3. **Fuse data.** Data fusion is the process of integrating multiple data items representing the same real-world object into a single, consistent, and clean representation. The main challenge in data fusion is the resolution of data conflicts, i.e., choosing a value in situations where multiple sources provide different values for the same property of an object. There is a large body of work on data fusion in the database community [35]. Linked Data applications can build on this work for choosing appropriate conflict resolution heuristics. Prototypical systems that support fusing Linked Data from multiple sources include DERI Pipes [73] and the KnoFuss architecture [89].

A list of criteria to assess the quality of Linked Data sources is proposed at [58]. In [56], Hartig presents an approach to handle trust (quality) values in SPARQL query processing. A method to integrate data quality assessment into query planning for federated architectures is presented in [85].

There is a large body of related work on probabilistic databases on which Linked Data applications can build. A survey of this work is presented in [45]. A well-known system which combines uncertainty and data linage is Trio System [2]. Uncertainty does not only apply to instance data but to a similar extent to vocabulary links that provide mappings between different terms. Existing work from databases that deal with the uncertainty of mappings in data integration processes is surveyed in [75].

Users will only trust the quality assessment results if they understand how these results were generated. Tim Berners-Lee proposed in [14] that Web browsers should be enhanced with an "Oh, yeah?" button to support the user in assessing the reliability of information encountered on the Web. Whenever a user encounters a piece of information that they would like to verify, pressing such a button would produce an explanation of the trustworthiness of the displayed information. This goal has yet to be realised; however, existing prototypes such as WIQA [28] and InferenceWeb [78]

---

[57]http://www4.wiwiss.fu-berlin.de/bizer/wiqa/
[58]http://sourceforge.net/apps/mediawiki/trdf/index.php?title=Quality_Criteria_for_Linked_Data_sources

provide explanations about information quality as well as inference processes and can be used as inspiration for work in this area.

6.3.6   CACHING WEB DATA LOCALLY

In order to query Web data more efficiently, it is often cached locally. The W3C Semantic Web Development Tools website[59] gives an overview of different RDF stores. The performance of these stores has improved considerably in recent years. Detailed benchmark results for the SPARQL query performance of different RDF stores are provided on the Berlin SPARQL Benchmark [33] Web site[60] as well as in the ESW wiki[61]. The current rule of thumb is that a single machine with enough RAM can store up to a billion RDF triples while providing SPARQL answers in a decent millisecond range. Rough estimates for clustered stores fall in a similar range, meaning that a reasonably powerful machine is needed for every one billion triples.

Standard SPARQL stores are optimized towards complex ad-hoc queries and are, of course, not the best choice for all types of queries. For example, most Linked Data search engines build their own indexing structures in order to provide for efficient low-expressivity queries over very large RDF data collections. For analytical tasks, there is a trend towards using MapReduce-based tools such as Hadoop[62].

6.3.7   USING WEB DATA IN THE APPLICATION CONTEXT

Web data that has been cached locally is usually either accessed via SPARQL queries or via an RDF API. A list of RDF APIs for different programming languages is maintained as part of the W3C Semantic Web Development Tools list[63]. An alternative list of Semantic Web development tools is *Sweet Tools*[64].

Well known RDF APIs include Jena[65] and Sesame[66]. A platform for developing Linked Data applications that provides a wide range of widgets for visualizing Web data is the Information Workbench[67].

6.4   EFFORT DISTRIBUTION BETWEEN PUBLISHERS, CONSUMERS AND THIRD PARTIES

Data integration is known to be a hard problem, and the problem is not getting easier as we aim to integrate large numbers of data sources in an open environment such as the Web. This section

[59]http://www.w3.org/2001/sw/wiki/Tools
[60]http://www4.wiwiss.fu-berlin.de/bizer/BerlinSPARQLBenchmark/
[61]http://esw.w3.org/RdfStoreBenchmarking
[62]http://hadoop.apache.org/mapreduce/
[63]http://www.w3.org/2001/sw/wiki/Tools
[64]http://www.mkbergman.com/new-version-sweet-tools-sem-web/
[65]http://jena.sourceforge.net/
[66]http://www.openrdf.org/doc/sesame/users/ch07.html
[67]http://iwb.fluidops.com/

describes how data integration is handled on the Web of Data. We will examine the interplay of integration efforts pursued by data publishers, data consumers and third parties and will show how they can complement each other to decrease heterogeneity in an evolutionary fashion over time.

In a classic data integration scenario, there may be three data sets provided by different parties. All use different data models and schemata but overlap in their content. In order to integrate these data sets, a data consumer will write some code or define some data integration workflow in an ETL environment. If a second data consumer wants to integrate the same data sets, she needs to go through exactly the same integration procedure again. Within this scenario, all data integration happens on the data consumer's side. Thus, the data consumer also needs to bear the complete effort of translating data between different schemata and detecting duplicate entities within the data sets.

This is different on the Web of Data as data publishers may contribute to making the integration easier for data consumers by reusing terms from widely used vocabularies, publishing mappings between terms from different vocabularies, and by setting RDF links pointing at related resources as well as at identifiers used by other data sources to refer to the same real-world entity. Therefore, we observe that the data integration effort is split between data providers and data consumers: the more effort data publishers put into publishing their data in a self-descriptive fashion [80], the easier it becomes for data consumers to integrate the data.

The Web of Data is a social system. Therefore, data publishers and data consumers are not the only players in a data integration scenario – there are third parties who may also contribute to the integration. Such third parties can be: vocabulary maintainers who publish vocabulary links relating terms from their vocabulary to other vocabularies; communities within industry or science that define mappings between the vocabularies that are commonly used in their domain of interest; or consumers of Linked Data that have invested effort into identity resolution or schema mapping and want to share the results of their efforts by making them available as identity or vocabulary RDF links on the Web.

Therefore, a major difference that distinguishes the Web of Data from other data integration environments is that the Web is used as a platform for sharing the results of integration efforts by different parties in the form of RDF links. This means that the overall data integration effort is split between these parties.

The downside of this open approach is that the quality of provided links is uncertain. Thus, the information consumer needs to interpret them as integration hints and make up her mind which ones she is willing to accept. Depending on the application domain and the number of data sets that should be integrated, this decision can be made manually (it is likely to be cheaper to review mappings than to generate them yourself) or based on data quality assessment heuristics such as the ones outlined in Section 6.3.5. An aspect that eases the task of the data consumer in verifying links as well as creating additional ones is that the Web of Data provides her with access to large amounts of instance data. Following Halevy's claim about the *unreasonable effectiveness of data* [52], the consumer can use simple cross-checking or voting techniques, or more advanced machine learning and data mining methods to verify integration hints and to learn new correspondences from the data [74].

It is also interesting to examine how heterogeneity on the Web of Data may decrease over time. Franklin, Halevy, and Maier have recognized that in large-scale integration scenarios involving thousands of data sources, it is impossible, or at least too expensive, to model a unifying integration schema upfront [51]. They have thus coined the term *dataspaces* for information systems that provide for the coexistence of heterogeneous data and do not require an upfront investment into a unifying schema. In such systems, data integration is achieved in a *pay-as-you-go* manner: as long as no or only a small number of mappings has been added to the system, applications can only display data in an unintegrated fashion and can only answer simple queries, or even only provide text search. Once more effort is invested over time in generating mappings, applications can further integrate the data and provide better query answers.

This is exactly what is happening at present on the Web of Data. In the absence of links and mappings, Linked Data applications, such as Tabulator, Marbles, Sindice or Falcons, display data in a rather unintegrated fashion. As more effort is invested over time into generating and publishing mappings on the Web, Linked Data applications can discover these mappings and use them to further integrate Web data and to deliver more sophisticated functionality. Thus, the Web of Data can be considered a dataspace according to Franklin, Halevy, and Maier's definition with the difference that the dataspace is distributed and of global scale.

The Web of Data relies on an evolutionary (i.e., tighter integration over time) as well as social (i.e., data publishers and third parties may contribute) data integration approach. In order to stress the social dimension of this integration process, we can speak of *somebody-pays-as-you-go integration* [27].

It will be interesting to see how vocabulary reuse [22], integration hints in the form of RDF links [27], and relationship mining [74] will play together over time in reaching the ultimate goal of Linked Data – being able to query the Web as a single global database.

CHAPTER 7

Summary and Outlook

This book has introduced the concept and basic principles of Linked Data, along with overviews of supporting technologies such as URIs, HTTP, and RDF. Collectively, these technologies and principles enable a style of publishing that weaves data into the very fabric of the Web – a unique characteristic that exposes Linked Data to the rigorous and boundless potential of the Web at large. This absolute integration with the Web, itself supported by open, community-driven standards, also serves a protective function for data publishers concerned about future-proofing their assets.

Linked Data has been adopted by a significant number of data publishers who, collectively, have built a Web of Data of impressive scale. In doing so, they have demonstrated its feasibility as an approach to publishing data in the Web and the growing maturity of the software platforms and toolsets that support this. Development of tools is always informed not just by standards and specifications, but by best practices that are developed and adopted in the relevant community. This book aims to reflect the best practices that have emerged since the first days of the Linking Open Data project, and record the best practices for the reference of others.

Perhaps the only aspect of Linked Data able to surpass the diversity of publishing scenarios will be the ways in which Linked Data is consumed from the Web. This book lays an initial foundation upon which different architectures for consuming Linked Data can be laid. Among this diversity, which is to be welcomed, there will inevitably be some common requirements and software components that reflect key characteristics and capabilities of Linked Data in the open, chaotic, and contradictory environment of the Web.

Large multi-national companies are currently facing similar challenges to those addressed in the Linked Data context: they maintain thousands of independently evolving databases across departments, subsidiaries and newly acquired companies, and struggle to realise the potential of their data assets. Therefore, just as classic Web technologies have been widely adopted within intranets, Linked Data has the potential to be used as light-weight, pay-as-you-go data integration technology within large organizations [114]. In contrast to classic data warehouses which require a costly upfront investment in modeling a global schema, Linked Data technologies allow companies to set up data spaces with relatively little effort. As these data spaces are being used, the companies can invest step-by-step in establishing data links, shared vocabularies, or schema mappings between the sources to allow deeper integration.

It is also interesting to note that major Web companies are already busy building such data spaces. Google, Yahoo and Facebook have all started to connect user, geographic and retail data, and begun to use these data spaces within their applications. In contrast to the open Web, which is accessible to everybody, these emerging data spaces are controlled by single companies which also decide how the data spaces are exploited – to the benefit of society as a whole or not [19]. Therefore, even though today it may be easier to enhance a Facebook profile or upload a database into Google Fusion Tables than to publish Linked Data, the effort may be well spent as it contributes to a public data asset able to as deeply influence the future of society as the arrival of the Web itself.

Bibliography

[1] Ben Adida and Mark Birbeck. Rdfa primer - bridging the human and data webs - w3c recommendation. http://www.w3.org/TR/xhtml-rdfa-primer/, 2008. 15, 18

[2] Parag Agrawal, Omar Benjelloun, Anish Das Sarma, Chris Hayworth, Shubha Nabar, Tomoe Sugihara, and Jennifer Widom. Trio: A system for data, uncertainty, and lineage. In *Proceedings of the VLDB Conference*, pages 1151–1154, 2006. 104

[3] Ali Al-Mahrubi and et al. http://dayta.me - A Personal News + Data Recommender for Your Day. Semantic Web Challenge 2010 Submission. http://www.cs.vu.nl/~pmika/swc/submissions/swc2010_submission_17.pdf, 2010. 91

[4] Keith Alexander. Rdf in json. In *Proceedings of the 4th Workshop on Scripting for the Semantic Web*, 2008. 20

[5] Keith Alexander, Richard Cyganiak, Michael Hausenblas, and Jun Zhao. Describing linked datasets. In *Proceedings of the WWW2009 Workshop on Linked Data on the Web*, 2009. 48

[6] Dean Allemang and Jim Hendler. *Semantic Web for the Working Ontologist: Effective Modeling in RDFS and OWL*. Morgan Kaufmann, 2008. 57, 63

[7] Christian Becker and Christian Bizer. Exploring the geospacial semantic web with dbpedia mobile. *Journal of Web Semantics: Science, Services and Agents on the World Wide Web*, 7:278–286, 2009. DOI: 10.1016/j.websem.2009.09.004 86, 92

[8] Christian Becker, Christian Bizer, Michael Erdmann, and Mark Greaves. Extending smw+ with a linked data integration framework. In *Proceedings of the ISWC 2010 Posters & Demonstrations Track*, 2010. 93

[9] D. Beckett. RDF/XML Syntax Specification (Revised) - W3C Recommendation. http://www.w3.org/TR/rdf-syntax-grammar/, 2004. 15, 18

[10] David Beckett and Tim Berners-Lee. Turtle - terse rdf triple language. http://www.w3.org/TeamSubmission/turtle/, 2008. 19

[11] Belleau, F., Nolin, M., Tourigny, N., Rigault, P., Morissette, J. Bio2rdf: Towards a mashup to build bioinformatics knowledge systems. *Journal of Biomedical Informatics*, 41(5):706–16, 2008. DOI: 10.1016/j.jbi.2008.03.004 37

[12] Michael Bergman. Advantages and myths of rdf. `http://www.mkbergman.com/wp-content/themes/ai3/files/2009Posts/Advantages_M%yths_RDF_090422.pdf`, 2009. 15

[13] Michael Bergman. What is a reference concept? `http://www.mkbergman.com/938/what-is-a-reference-concept/`, 2010. 65

[14] Tim Berners-Lee. Cleaning up the User Interface, 1997. `http://www.w3.org/DesignIssues/UI.html`. 104

[15] Tim Berners-Lee. Cool uris don't change. `http://www.w3.org/Provider/Style/URI`, 1998. 43

[16] Tim Berners-Lee. Linked Data - Design Issues, 2006. `http://www.w3.org/DesignIssues/LinkedData.html`. 7, 26, 82

[17] Tim Berners-Lee. Giant global graph. `http://dig.csail.mit.edu/breadcrumbs/node/215`, 2007. 16, 29

[18] Tim Berners-Lee. Putting government data online. `http://www.w3.org/DesignIssues/GovData.html`, 2009. 36

[19] Tim Berners-Lee. Long live the web: A call for continued open standards and neutrality. *Scientific American*, 32, 2010. 110

[20] Tim Berners-Lee, R. Fielding, and L. Masinter. *RFC 2396 - Uniform Resource Identifiers (URI): Generic Syntax*. `http://www.isi.edu/in-notes/rfc2396.txt`, August 1998. 7

[21] Tim Berners-Lee, James Hendler, and Ora Lassilia. The semantic web. *Scientific American*, 284(5):34–44, Mai 2001. DOI: 10.1038/scientificamerican0501-34 5

[22] Tim Berners-Lee and Lalana Kagal. The fractal nature of the semantic web. *AI Magazine*, Vol 29, No 3, 2008. 24, 62, 101, 107

[23] Diego Berrueta and Jon Phipps. Best practice recipes for publishing rdf vocabularies - w3c note. `http://www.w3.org/TR/swbp-vocab-pub/`, 2008. 24, 58, 63, 83

[24] Alexander Bilke and Felix Naumann. Schema matching using duplicates. In *Proceedings of the International Conference on Data Engineering*, 2005. DOI: 10.1109/ICDE.2005.126 102

[25] Mark Birbeck. Rdfa and linked data in uk government web-sites. *Nodalities Magazine*, 7, 2009. 36

[26] Paul Biron and Ashok Malhotra. Xml schema part 2: Datatypes second edition - w3c recommendation. `http://www.w3.org/TR/xmlschema-2/`, 2004. 16

[27] Christian Bizer. Pay-as-you-go data integration on the public web of linked data. Invited talk at the 3rd Future Internet Symposium 2010. http://www.wiwiss.fu-berlin.de/en/institute/pwo/bizer/research/publications/%Bizer-FIS2010-Pay-As-You-Go-Talk.pdf, 2010. 107

[28] Christian Bizer and Richard Cyganiak. Quality-driven information filtering using the wiqa policy framework. *Journal of Web Semantics: Science, Services and Agents on the World Wide Web*, 7(1):1–10, 2009. DOI: 10.1016/j.websem.2008.02.005 103, 104

[29] Christian Bizer, Richard Cyganiak, and Tobias Gauss. The rdf book mashup: From web apis to a web of data. In *Proceedings of the Workshop on Scripting for the Semantic Web*, 2007. 38, 70

[30] Christian Bizer, Tom Heath, and Tim Berners-Lee. Linked data - the story so far. *Int. J. Semantic Web Inf. Syst.*, 5(3):1–22, 2009. DOI: 10.4018/jswis.2009081901 5, 29

[31] Christian Bizer, Ralf Heese, Malgorzata Mochol, Radoslaw Oldakowski, Robert Tolksdorf, and Rainer Eckstein. The impact of semantic web technologies on job recruitment processes. In *Proceedings of the 7. Internationale Tagung Wirtschaftsinformatik (WI2005)*, 2005. DOI: 10.1007/3-7908-1624-8_72 36

[32] Christian Bizer, Jens Lehmann, Georgi Kobilarov, Sören Auer, Christian Becker, Richard Cyganiak, and Sebastian Hellmann. Dbpedia - a crystallization point for the web of data. *Journal of Web Semantics: Science, Services and Agents on the World Wide Web*, 7(3):154–165, 2009. DOI: 10.1016/j.websem.2009.07.002 14, 33

[33] Christian Bizer and Andreas Schultz. The berlin sparql benchmark. *International Journal on Semantic Web and Information Systems*, 5(2):1–24, 2009. 105

[34] Christian Bizer and Andreas Schultz. The r2r framework: Publishing and discovering mappings on the web. In *Proceedings of the 1st International Workshop on Consuming Linked Data*, 2010. 25, 102

[35] Bleiholder, J., Naumann, F. Data fusion. *ACM Computing Surveys*, 41(1):1–41, 2008. DOI: 10.1145/1456650.1456651 104

[36] John Breslin, Andreas Harth, Uldis Bojars, and Stefan Decker. Towards semantically-interlinked online communities. In *Proceedings of the 2nd European Semantic Web Conference*, Heraklion, Greece, 2005. DOI: 10.1007/11431053_34 54

[37] D. Brickley and R. V. Guha. RDF Vocabulary Description Language 1.0: RDF Schema - W3C Recommendation. http://www.w3.org/TR/rdf-schema/, 2004. 17, 24, 56

[38] Peter Buneman, Sanjeev Khanna, and Wang chiew Tan. Why and where: A characterization of data provenance. In *Proceedings of the International Conference on Database Theory*, pages 316–330. Springer, 2001. DOI: 10.1007/3-540-44503-X_20 52

[39] Carroll, J., Bizer, C., Hayes, P., Stickler, P. Named graphs. *Journal of Web Se-mantics: Science, Services and Agents on the World Wide Web*, 3(4):247–267, 2005. DOI: 10.1016/j.websem.2005.09.001 95, 103

[40] Gong Cheng and Yuzhong Qu. Searching linked objects with falcons: Approach, implementation and evaluation. *International Journal on Semantic Web and Information Systems (IJSWIS)*, 5(3):49–70, 2009. 87

[41] Chris Clarke. A resource list management tool for undergraduate students based on linked open data principles. In *Proceedings of the 6th European Semantic Web Conference*, Heraklion, Greece, 2009. DOI: 10.1007/978-3-642-02121-3_51 91

[42] Gianluca Correndo, Manuel Salvadores, Ian Millard, Hugh Glaser, and Nigel Shadbolt. Sparql query rewriting for implementing data integration over linked data. In *Proceedings of the 2010 EDBT/ICDT Workshops*, pages 1–11, New York, NY, USA, 2010. ACM. DOI: 10.1145/1754239.1754244 102

[43] Cyganiak, R., Delbru, R., Stenzhorn, H., Tummarello, G., Decker, S. Semantic sitemaps: Efficient and flexible access to datasets on the semantic web. In *Proceedings of the 5th European Semantic Web Conference*, 2008. DOI: 10.1007/978-3-540-68234-9_50 48

[44] Axel Cyrille, Ngonga Ngomo, and Sören Auer. Limes - a time-efficient approach for large-scale link discovery on the web of data. `http://svn.aksw.org/papers/2011/WWW_LIMES/public.pdf`, 2010. 68

[45] Nilesh Dalvi, Christopher Ré, and Dan Suciu. Probabilistic databases: diamonds in the dirt. *Commun. ACM*, 52:86–94, July 2009. DOI: 10.1145/1538788.1538810 104

[46] Elmagarmid, A., Ipeirotis, P., Verykios, V. . Duplicate record detection: A survey. *IEEE Transactions on Knowledge and Data Engineering*, 19(1):1–16, 2007. DOI: 10.1109/TKDE.2007.250581 66, 102

[47] Jim Ericson. Net expectations - what a web data service economy implies for business. *Information Management Magazine*, Jan/Feb, 2010. 1

[48] Euzenat, J., Scharffe, F., Zimmermann A. Expressive alignment language and implementation. Knowledge Web project report, KWEB/2004/D2.2.10/1.0, 2007. 102

[49] Euzenat, J., Shvaiko, P. *Ontology Matching*. Springer, Heidelberg, 2007. 66, 102

[50] Roy Fielding. Hypertext transfer protocol – http/1.1. request for comments: 2616. `http://www.w3.org/Protocols/rfc2616/rfc2616.html`, 1999. 7, 10

[51] Franklin, M.J., Halevy, A.Y., Maier, D. From databases to dataspaces: A new abstraction for information management. *SIGMOD Record*, 34(4):27–33, 2005. DOI: 10.1145/1107499.1107502 25, 107

[52] Alon Y. Halevy, Peter Norvig, and Fernando Pereira. The unreasonable effectiveness of data. *IEEE Intelligent Systems*, 24(2):8–12, 2009. DOI: 10.1109/MIS.2009.36 106

[53] Harry Halpin, Patrick Hayes, James McCusker, Deborah Mcguinness, and Henry Thompson. When owl:sameas isn't the same: An analysis of identity in linked data. In *Proceedings of the 9th International Semantic Web Conference*, 2010. DOI: 10.1007/978-3-642-17746-0_20 23

[54] Andreas Harth. Visinav: A system for visual search and navigation on web data. *Web Semantics: Science, Services and Agents on the World Wide Web*, 8(4):348 – 354, 2010. DOI: 10.1007/978-3-642-03573-9_17 89

[55] Andreas Harth, Aidan Hogan, Jürgen Umbrich, and Stefan Decker. Swse: Objects before documents! In *Proceedings of the Semantic Web Challenge 2008*, 2008. 87

[56] Olaf Hartig. Querying trust in rdf data with tsparql. In *Proceedings of the 6th European Semantic Web Conference*, pages 5–20, 2009. DOI: 10.1007/978-3-642-02121-3_5 104

[57] Olaf Hartig, Christian Bizer, and Johann Christoph Freytag. Executing sparql queries over the web of linked data. In *Proceedings of the International Semantic Web Conference*, pages 293–309, 2009. DOI: 10.1007/978-3-642-04930-9_19 97, 100

[58] Olaf Hartig and Andreas Langegger. A database perspective on consuming linked data on the web. *Datenbank-Spektrum*, 10:57–66, 2010. 10.1007/s13222-010-0021-7. DOI: 10.1007/s13222-010-0021-7 98

[59] Olaf Hartig, Hannes Mühleisen, and Johann-Christoph Freytag. Linked data for building a map of researchers. In *Proceedings of the 5th Workshop on Scripting for the Semantic Web*, 2009. 93

[60] Haslhofer, B. *A Web-based Mapping Technique for Establishing Metadata Interoperability*. PhD thesis, Universitaet Wien, 2008. 102

[61] Tom Heath and Enrico Motta. Revyu: Linking reviews and ratings into the web of data. *Journal of Web Semantics: Science, Services and Agents on the World Wide Web*, 6(4), 2008. DOI: 10.1016/j.websem.2008.09.003 38

[62] Heath, T.. How will we interact with the web of data? *IEEE Internet Computing*, 12(5):88–91, 2008. DOI: 10.1109/MIC.2008.101 86

[63] Martin Hepp. Goodrelations: An ontology for describing products and services offers on the web. In *Proceedings of the 16th International Conference on Knowledge Engineering and Knowledge Management*, Acitrezza, Italy, 2008. 38

[64] Aidan Hogan, Andreas Harth, Alexandre Passant, Stefan Decker, and Axel Polleres. Weaving the pedantic web. In *Proceedings of the WWW2010 Workshop on Linked Data on the Web*, 2010. 82

[65] Robert Isele, Andreas Harth, Jürgen Umbrich, and Christian Bizer. Ldspider: An open-source crawling framework for the web of linked data. In *ISWC 2010 Posters & Demonstrations Track: Collected Abstracts Vol-658*, 2010. 94, 100

[66] Robert Isele, Anja Jentzsch, and Christian Bizer. Silk server - adding missing links while consuming linked data. In *Proceedings of the 1st International Workshop on Consuming Linked Data (COLD 2010)*, 2010. 102

[67] Ian Jacobs and Norman Walsh. *Architecture of the World Wide Web, Volume One*, 2004. http://www.w3.org/TR/webarch/. 7, 9

[68] Anja Jentzsch, Oktie Hassanzadeh, Christian Bizer, Bo Andersson, and Susie Stephens. Enabling tailored therapeutics with linked data. In *Proceedings of the WWW2009 Workshop on Linked Data on the Web*, 2009. 5, 37

[69] Clement Jonquet, Paea LePendu, Sean Falconer, Adrien Coulet, Natalya Noy, Mark Musen, and Nigam Shah. Ncbo resource index: Ontology-based search and mining of biomedical resources. Semantic Web Challenge 2010 Submission. http://www.cs.vu.nl/~pmika/swc/submissions/swc2010_submission_4.pdf, 2010. 92

[70] Graham Klyne and Jeremy J. Carroll. *Resource Description Framework (RDF): Concepts and Abstract Syntax - W3C Recommendation*, 2004. http://www.w3.org/TR/rdf-concepts/. 8, 15, 17

[71] Georgi Kobilarov, Tom Scott, Yves Raimond, Silver Oliver, Chris Sizemore, Michael Smethurst, Christian Bizer, and Robert Lee. Media meets semantic web - how the bbc uses dbpedia and linked data to make connections. In *The Semantic Web: Research and Applications, 6th European Semantic Web Conference*, pages 723–737, 2009. DOI: 10.1007/978-3-642-02121-3_53 34

[72] Donald Kossmann. The state of the art in distributed query processing. *ACM Comput. Surv.*, 32:422–469, December 2000. DOI: 10.1145/371578.371598 98

[73] Danh Le-Phuoc, Axel Polleres, Manfred Hauswirth, Giovanni Tummarello, and Christian Morbidoni. Rapid prototyping of semantic mash-ups through semantic web pipes. In *Proceedings of the 18th international conference on World wide web*, WWW '09, pages 581–590. ACM, 2009. DOI: 10.1145/1526709.1526788 104

[74] Madhavan, J., Shawn, J. R., Cohen, S., Dong, X., Ko, D., Yu, C., Halevy, A. Web-scale data integration: You can only afford to pay as you go. *Proceedings of the Conference on Innovative Data Systems Research*, 2007. 25, 106, 107

[75] Matteo Magnani and Danilo Montesi. A survey on uncertainty management in data integration. *J. Data and Information Quality*, 2:5:1–5:33, July 2010. DOI: 10.1145/1805286.1805291 104

[76] Frank Manola and Eric Miller. *RDF Primer*. W3C, `http://www.w3c.org/TR/rdf-primer/`, February 2004. 15, 18

[77] Philippe C. Mauroux, Parisa Haghani, Michael Jost, Karl Aberer, and Hermann De Meer. idMesh: graph-based disambiguation of linked data. In *Proceedings of the 18th international conference on World wide web*, pages 591–600. ACM, 2009. DOI: 10.1145/1526709.1526789 68

[78] Deborah L. McGuinness and Paulo Pinheiro da Silva. Inference web: Portable and shareable explanations for question answering. In *Proceedings of the American Association for Artificial Intelligence Spring Symposium Workshop on New Directions for Question Answering. Stanford University*, 2003. 104

[79] Deborah L. McGuinness and Frank van Harmelen. OWL Web Ontology Language Overview - W3C Recommendation. `http://www.w3.org/TR/2004/REC-owl-features-20040210/`, 2004. 17, 24, 56

[80] Noah Mendelsohn. The self-describing web - tag finding. `http://www.w3.org/2001/tag/doc/selfDescribingDocuments.html`, 2009. 24, 29, 106

[81] Alistair Miles and Sean Bechhofer. Skos simple knowledge organization system - reference. `http://www.w3.org/TR/skos-reference/`, 2009. 24, 56

[82] Paul Miller, Rob Styles, and Tom Heath. Open data commons, a license for open data. In *Proceedings of the WWW2008 Workshop on Linked Data on the Web*, 2008. 53

[83] R. Moats. Rfc 2141: Urn syntax. `http://tools.ietf.org/html/rfc2141`, 1997. 10

[84] Knud Möller, Tom Heath, Siegfried Handschuh, and John Domingue. Recipes for semantic web dog food - the eswc and iswc metadata projects. In *Proceedings of the 6th International Semantic Web Conference and 2nd Asian Semantic Web Conference*, Busan, Korea, 2007. DOI: 10.1007/978-3-540-76298-0_58 37

[85] Felix Naumann. *Quality-Driven Query Answering for Integrated Information Systems*, volume 2261 / 2002 of *Lecture Notes in Computer Science*. Springer-Verlag GmbH, 2002. 104

[86] Joachim Neubert. Bringing the "thesaurus for economics" on to the web of linked data. In *Proceedings of the WWW2009 Workshop on Linked Data on the Web*, 2009. 36

[87] Popitsch Niko and Haslhofer Bernhard. Dsnotify: Handling broken links in the web of data. In *Proceedings of the 19th International World Wide Web Conference*, Raleigh, NC, USA, 2 2010. ACM. DOI: 10.1145/1772690.1772768 68

[88] Andriy Nikolov and Enrico Motta. Capturing emerging relations between schema ontologies on the web of data. In *Proceedings of the First International Workshop on Consuming Linked Data*, 2010. 102

[89] Andriy Nikolov, Victoria Uren, Enrico Motta, and Anne Roeck. Integration of seman-tically annotated data by the knofuss architecture. In *Proceedings of the 16th interna-tional conference on Knowledge Engineering: Practice and Patterns*, pages 265–274, 2008. DOI: 10.1007/978-3-540-87696-0_24 104

[90] Benjamin Nowack. Paggr: Linked data widgets and dashboards. *Web Semantics: Science, Services and Agents on the World Wide Web*, 7(4):272 – 277, 2009. Semantic Web challenge 2008. 91

[91] Lawrence Page, Sergey Brin, Rajeev Motwani, and Terry Winograd. The pagerank citation ranking: Bringing order to the web. Technical report, Stanford Digital Library Technologies Project, 1998. 104

[92] Norman Paskin. The doi handbook - edition 4.4.1. http://www.doi.org/handbook_2000/DOIHandbook-v4-4.1.pdf, 2006. 10

[93] P. F. Patel-Schneider, P. Hayes, and I. Horrocks. OWL Web Ontology Language Se-mantics and Abstract Syntax - W3C Recommendation. http://www.w3.org/TR/owl-semantics/, 2004. 23

[94] Axel Polleres, François Scharffe, and Roman Schindlauer. Sparql++ for mapping between rdf vocabularies. In *Proceedings of the 6th International Conference on Ontologies, DataBases, and Applications of Semantics (ODBASE 2007)*, 2007. 102

[95] Eric Prud'hommeaux and Andy Seaborne. *SPARQL Query Language for RDF - W3C Rec-ommendation*, 2008. http://www.w3.org/TR/rdf-sparql-query/. 17, 96

[96] Bastian Quilitz and Ulf Leser. Querying distributed rdf data sources with sparql. In *Proceedings of the 5th European Semantic Web Conference*, 2008. DOI: 10.1007/978-3-540-68234-9_39 101

[97] Dave Raggett, Arnaud Le Hors, and Ian Jacobs. Html 4.01 specification - w3c recommen-dation. http://www.w3.org/TR/html401/, 1999. 7

[98] Leo Sauermann and Richard Cyganiak. Cool uris for the semantic web - w3c interest group note. http://www.w3.org/TR/cooluris/, 2008. 11, 13, 14, 15, 43, 46

[99] Simon Schenk, Carsten Saathoff, Steffen Staab, and Ansgar Scherp. Semaplorer–interactive semantic exploration of data and media based on a federated cloud infrastructure. *Web Se-mantics: Science, Services and Agents on the World Wide Web*, 7(4):298 – 304, 2009. Semantic Web challenge 2008. 101

[100] Len Seligman, Peter Mork, Alon Y. Halevy, Ken Smith, Michael J. Carey, Kuang Chen, Chris Wolf, Jayant Madhavan, Akshay Kannan, and Doug Burdick. Openii: an open source

information integration toolkit. In *Proceedings of the SIGMOD Conference*, pages 1057–1060, 2010. DOI: 10.1145/1807167.1807285 102

[101] John Sheridan and Jeni Tennison. Linking uk government data. In *Proceedings of the WWW2010 Workshop on Linked Data on the Web*, 2010. 36

[102] Sebastian Hellmann Sören Auer, Jens Lehmann. Linkedgeodata - adding a spatial dimension to the web of data. In *Proceedings of the International Semantic Web Conference*, 2009. DOI: 10.1007/978-3-642-04930-9_46 34

[103] Rob Styles, Danny Ayers, and Nadeem Shabir. Semantic marc, marc21 and the semantic web. In *Proceedings of the WWW2008 Workshop on Linked Data on the Web*, 2008. 43

[104] Fabian M. Suchanek, Gjergji Kasneci, and Gerhard Weikum. Yago: a core of semantic knowledge. In Carey L. Williamson, Mary Ellen Zurko, Peter F. Patel-Schneider, and Prashant J. Shenoy, editors, *Proceedings of the 16th International Conference on World Wide Web, WWW 2007, Banff, Alberta, Canada, May 8-12, 2007*, pages 697–706. ACM, 2007. 34

[105] T. Berners-Lee et al. Tabulator: Exploring and analyzing linked data on the semantic web. In *Proceedings of the 3rd International Semantic Web User Interaction Workshop*, 2006. 86

[106] Henry Thompson and David Orchard. Urns, namespaces and registries. `http://www.w3.org/2001/tag/doc/URNsAndRegistries-50`, 2006. 10

[107] Giovanni Tummarello, Richard Cyganiak, Michele Catasta, Szymon Danielczyk, Renaud Delbru, and Stefan Decker. Sig.ma: Live views on the web of data. *Web Semantics: Science, Services and Agents on the World Wide Web*, 8(4):355 – 364, 2010. DOI: 10.1145/1772690.1772907 87

[108] Giovanni Tummarello, Renaud Delbru, and Eyal Oren. Sindice.com: Weaving the Open Linked Data. In *Proceedings of the 6th International Semantic Web Conference*, 2007. 89

[109] Jürgen Umbrich, Boris Villazon-Terrazas, and Michael Hausenblas. Dataset dynamics compendium: A comparative study. In *Proceedings of the First International Workshop on Consuming Linked Data (COLD2010)*, 2010. 68

[110] Van de Sompel, H., Lagoze, C., Nelson, M., Warner, S., Sanderson, R., Johnston, P. Adding escience assets to the data web. In *Proceedings of the 2nd Workshop on Linked Data on the Web (LDOW2009)*, 2009. 37

[111] Julius Volz, Christian Bizer, Martin Gaedke, and Georgi Kobilarov. Discovering and maintaining links on the web of data. In *Proceedings of the International Semantic Web Conference*, pages 650–665, 2009. DOI: 10.1007/978-3-642-04930-9_41 68

[112] Denny Vrandecic, Varun Ratnakar, Markus Krötzsch, and Yolanda Gil. Shortipedia - aggregating and curating semantic web data. Semantic Web Challenge 2010 Submission. `http://www.cs.vu.nl/~pmika/swc/submissions/swc2010_submission_18.pdf`, 2010. 93

[113] Norman Walsh. Names and Addresses. `http://norman.walsh.name/2006/07/25/namesAndAddresses`, 2006. 10

[114] David Wood. *Linking Enterprise Data*. Springer, 2010. 5, 109

Authors' Biographies

DR. TOM HEATH

Dr. Tom Heath is lead researcher at Talis, a Birmingham, UK-based software company and global leader in the research, development and commercial exploitation of Linked Data and Semantic Web technologies. At Talis, he is responsible for leading internal research exploring how Linked Data affects the sharing and reuse of data, the value and insights that can be derived from this data, and the implications of these changes for human-computer interaction. Tom has a first degree in psychology from the University of Liverpool and a PhD in Computer Science from The Open University's Knowledge Media Institute. His doctoral research examined how a richer understanding of trust decisions in word-of-mouth recommendation can be combined with Semantic Web technologies and social networks to improve the personal relevance of information-seeking processes on the Web.

Tom has been active in the Linked Data community since its inception in early 2007, creating the Linked Data reviewing and rating site Revyu.com that went on to win first prize in the 2007 International Semantic Web Challenge. In addition to co-founding the highly successful workshop series *Linked Data on the Web*, held at the International World-wide Web Conference since 2008, he has served as chair of the Semantic Web In-use track at both the International and European Semantic Web Conferences, and he co-authored the seminal article *Linked Data - The Story So Far* with Christian Bizer and Tim Berners-Lee. In 2009, Tom was named *PhD of the Year* 2008/9 by STI International, and in 2011 one of *AI's 10 to Watch* by IEEE Intelligent Systems.

PROF. DR. CHRISTIAN BIZER

Prof. Dr. Christian Bizer is the head of the Web-based Systems Group at Freie Universität Berlin, Germany. He explores technical and economic questions concerning the development of global, decentralized information environments. The current focuses of his research are data integration, identity resolution, and information quality assessment in the context of the Web of Data. The results of his work include the *Named Graphs* data model which was adopted into the W3C SPARQL recommendation; the *D2RQ* mapping language, which is widely used for publishing relational databases to the Web of Data; the *Fresnel* display vocabulary implemented by several Linked Data browsers, and the *Berlin SPARQL Benchmark* for measuring the performance of RDF stores. He is one of the co-founders of the *W3C Linking Open Data* community effort which aims at interlinking large numbers of data sources on the Web. He also co-founded the *DBpedia* project which extracts a comprehensive knowledge base from Wikipedia and has developed into an interlinking hub in the Web of Data. Christian Bizer is chairing the *Semantic Web Challenge* series at the International Semantic Web Conference and is one of the co-founders of the *Linked Data on the Web* workshop series. He obtained his doctoral degree with a dissertation on information quality in the context of Web-based Systems and has published over 30 papers in journals, conference and workshop proceedings.

Printed in the United States
by Baker & Taylor Publisher Services